T0197821

The Water Recycling Revolution

The Water Recycling Revolution

Tapping into the Future

William M. Alley
and
Rosemarie Alley

ROWMAN & LITTLEFIELD
Lanham • Boulder • New York • London

Published by Rowman & Littlefield
An imprint of The Rowman & Littlefield Publishing Group, Inc.
4501 Forbes Boulevard, Suite 200, Lanham, Maryland 20706
www.rowman.com

86-90 Paul Street, London EC2A 4NE, United Kingdom

British Library Cataloguing in Publication Information Available

Library of Congress Cataloging-in-Publication Data

Names: Alley, William M., author. | Alley, Rosemarie, author.
Title: The water recycling revolution : tapping into the future / William M. Alley
   and Rosemarie Alley.
Description: Lanham : Rowman & Littlefield, 2022. | Includes bibliographical
   references and index. | Summary: "Tells the story of recycled wastewater,
   examines the pros and cons, and explores its future potential"— Provided
   by publisher.
Identifiers: LCCN 2021040393 (print) | LCCN 2021040394 (ebook) | ISBN
   9781538160411 (cloth) | ISBN 9781538160428 (epub)
Subjects: LCSH: Water reuse—United States. | Sewage—Purification—United
   States. | Water—Purification—United States.
Classification: LCC TD429 .A425 2022 (print) | LCC TD429 (ebook) | DDC
   628.1/62—dc23
LC record available at https://lccn.loc.gov/2021040393
LC ebook record available at https://lccn.loc.gov/2021040394

To the many dedicated public servants who daily
safeguard our essential water services.

Contents

Contents

Acknowledgments

We greatly benefited from discussions with many individuals who generously gave of their time to share their knowledge and insights about water reuse. Key among these are Greg Baker, Laura Belanger, Robert Beltran, Brian Biesemeyer, Taylor Chang, Dave Colvin, Jason Dadakis, Lisa Darling, John Fabris, Anthony Fellow, Marco Gonzalez, Earle Hartling, Todd Hartman, Ted Henifin, Roy Herndon, Wayne Hill, Adam Hutchinson, Ted Johnson, Paula Kehoe, Andy Kricun, Keith Lewinger, Richard Luthy, Mark Marlowe, Melissa McChesney, Melissa Meeker, Mark Millan, Christina Montoya, Seung Park, Austa Parker, Jim Poff, Gayathri RamMohan, Bruce Reznik, Sybil Sharvelle, Bahman Sheikh, Ann Shortelle, Marsi Steirer, Patricia Tennyson, Mark Tettemer, Bruce Thomson, Gilbert Trejo, Shane Trussell, William Van Wagoner, Chuck Weber, and Bart Weiss. At Rowman & Littlefield, we are grateful to Suzanne Staszak-Silva, who was an enthusiastic supporter of the project from the start. We also thank Brianna Westervelt and Brennan Knight for their support during the production process. The perspectives on water reuse and any errors are ours alone.

Introduction

Every day in cities across the country, billions of gallons of municipal wastewater (aka sewage) are treated and discharged into a nearby waterbody. Some of the wastewater is recycled. But, in the scheme of things, the total amount is just a drop in the bucket. Less than 1 percent of U.S. water demand is currently being met through water reuse. There's plenty of room for expansion as urban population growth, environmental needs, and climate change pressure drinking water supplies of cities across the country, as well as worldwide.

While nonpotable uses, such as landscape irrigation, are mostly noncontroversial, it's a different story for recycling wastewater for potable uses. From the earliest human settlements, there has been a taboo against mixing sewage and drinking water. This automatic reaction runs so deep that it's essentially in our DNA. The discovery that waterborne pathogens cause diseases like cholera upped the ante.

In recent years, humans have begun to turn this age-old taboo on its head by using advanced-treated wastewater to supplement a city's drinking water supply. This increasingly widespread practice, known as potable reuse, qualifies as nothing less than a drinking water revolution. In this book, we track the story of this development, examine the pros and cons, and explore its future potential. Nonpotable reuse is addressed along the way.

Any number of challenges and obvious questions arise when a water utility is exploring the possibility of potable reuse. How do you get people to overcome the visceral reaction known as the "Yuck Factor" and not only drink, but also appreciate, recycled wastewater? What motivates water utilities to consider potable reuse, which not long ago was considered the solution of last resort? Which is better: *Direct* potable reuse that connects recycled wastewater directly to the drinking water system, or *indirect* potable reuse that releases purified water to an environmental buffer, such as a lake or groundwater, from which withdrawals are treated to meet potable water standards? And what about all those pharmaceuticals and personal care products that people casually flush down the drain, or that are shed from our bodies?

There are also less obvious questions. Will diverting discharges from a wastewater-treatment plant damage downstream users or ecosystems that previously depended on that water? What about water rights? Why not just use the treated wastewater for nonpotable uses, such as irrigating lawns and parks? For coastal cities, why not just turn to desalination of the virtually unlimited ocean water at their doorstep? What are the implications for climate change and drought mitigation?

We explore these and other questions as we investigate a variety of places across the country that have turned to potable reuse. Some are obvious, such as California, Arizona, and Texas. Others may come as a surprise, such as Georgia, Virginia, and a small mountain town in New Mexico. Each has a unique story of what led them to develop potable reuse for securing their drinking water supply—as well as the challenges they had to overcome. A commonality for nearly all of them was selling their customers on the idea.

The issue is widely relevant. In the same way that the electric car is changing the future of transportation, potable reuse is part of the sustainable city of the future in which we rethink "waste" water as a renewable resource. Depending on where you live, work, or travel, when you flush the toilet or take your next drink of water, you may now or sometime in the near future be participating in this water-recycling revolution.

WASTEWATER TREATMENT 101

For those new to this topic, let's take a brief look at some basic wastewater treatment concepts and terminology (those familiar with wastewater treatment may want to go directly to chapter 1). First, unless we note otherwise, *wastewater* refers to the municipal wastewater that comes from households, businesses, hospitals, and industrial facilities hooked up to the sewer system. Notably and contrary to common view, human waste is a small percentage of this wastewater.

Conventional wastewater treatment is relatively simple in concept. In a preliminary step, grates and screens catch large materials and grit as wastewater enters the treatment facility. All kinds of things find their way into wastewater—rags, toys, golf balls, you name it. So-called "flushable wipes" have been a particular bane. After the preliminary screening come primary and secondary treatments.

Primary treatment physically separates waste from water. As wastewater slowly moves through sedimentation tanks, heavier material settles to the bottom of the tank. A giant skimming arm scrapes away fats, oil, and grease that float along the surface of the water.

Secondary treatment takes advantage of biological processes. The most common method is known by the somewhat cryptic and unappealing name of "activated sludge." Bacteria are added and air is bubbled up through the effluent, speeding up what nature does as water flows over rocks and ledges. These good bacteria (aka "bugs") gobble up the organic matter and help to settle out more of the suspended solids. Heavy metals and other contaminants that may be attached to the suspended solids are also removed. After consuming the organic material, some of the microorganisms are settled out in clarifiers and reintroduced to consume more of the organic materials. Other methods, such as trickling filter beds, act similarly to activated sludge in using microorganisms to remove organics under aerobic conditions.

Secondary treatment is capable of removing over 90 percent of the suspended solids and oxygen-demanding wastes. As such, it addresses many of the problems of aesthetics and die-off of fish and aquatic life that once plagued rivers, streams, and lakes throughout the country.

The final process in conventional wastewater treatment is disinfection to kill or inactivate harmful microorganisms still left in the water. This is typically done through chlorine, ozone, or ultraviolet disinfection (or a combination of these). Chlorine destroys target organisms through a complex process that involves oxidizing their cellular material. Treatment plant operators will generally remove the chlorine before releasing the effluent, so the chlorine doesn't damage the environment. Ozone, commonly formed using electricity to cause oxygen gas molecules to turn into ozone molecules, destroys the cell wall of the microorganisms and kills them. Ultraviolet light (UV) scrambles the DNA of the micro-organisms, making it impossible for them to reproduce.

For discharge to environmentally sensitive waterways, additional *tertiary treatment* may be required. This can take several different forms and often involves some type of filtration followed by disinfection. Tertiary treatment is often used to remove nitrogen and phosphorus prior to discharge of treated effluent to waters vulnerable to algal blooms and other effects of excess nutrients.

The additional steps beyond tertiary treatment are referred to as *advanced wastewater treatment, advanced water treatment,* or *advanced water purification*. The latter term is increasingly popular, as studies have shown that *purification* resonates positively with the public. We use these terms interchangeably. Water that has been through processes beyond tertiary treatment is referred to as *advanced-treated water*.

There are many possible *treatment trains*—series of treatment steps—that can be applied for advanced wastewater treatment. These include various combinations of microfiltration, ultrafiltration, nano-filtration, membrane bioreactors, reverse osmosis, granular activated carbon, biologically activated carbon, UV light, advanced oxidation processes, and so forth—you get the idea. We'll cover many of these techniques briefly as they arise in our examples. Our purpose, however, is not to go into the details on individual treatment techniques, as there are many other sources of that information. As we'll see, two treatment combinations commonly included in treatment trains are (1) micro-filtration–reverse osmosis–UV/advanced oxidation processes, and (2) ozone–biologically activated carbon often followed by exposure to UV light.

In addition to the techniques used in wastewater-treatment plants, various natural cleansing processes are commonly employed as a key component of advanced wastewater treatment to enhance water quality and provide natural storage for reclaimed water. These fall into three general categories: managed aquifer recharge (the purposeful recharge of water to aquifers), surface water augmentation using a reservoir or stream reach, and use of natural and constructed wetlands. We'll explore examples of each.

There's another type of potable reuse that has existed throughout modern history, but people don't often think about it that way. *De facto potable reuse* occurs where a substantial portion of the drinking water drawn from a stream originated as wastewater effluent from upstream communities. De facto potable reuse is not officially recognized as water reuse, but during low flows, many streams contain a large fraction of wastewater. Fortunately, nature helps clean up the water on its way to the drinking water intake, but it may still contain pharmaceuticals and other chemicals resistant to breakdown, as well as pathogens.

Finally, a few words about the terms *water reuse, water recycling,* and *water reclamation. Water reuse* and *water recycling* are almost always used interchangeably. *Water reclamation* often refers to water that has been treated but not yet reused. We use all three terms, while noting that *recycling* and *reuse* have greater public appeal than *reclamation* or *reclaimed water.*

Chapter One

America's Finest City

Civilized people should be able to dispose of sewage in a better way than by putting it in the drinking water.

—Theodore Roosevelt[1]

San Diego's wastewater-treatment plant is a well-kept secret, hidden away as it is on Point Loma, a finger-like peninsula less than a half-mile wide that juts out into the Pacific Ocean. Out of sight and out of mind, the plant pumps around 175 million gallons a day of partially treated sewage to an ocean outfall over four miles offshore and 320 feet beneath the ocean surface.

The situation is a major improvement over the days when untreated sewage was dumped into San Diego Bay, corroding the hulls of Navy ships and driving tourists away.[2] Although the bay recovered after the Point Loma plant opened in 1963, the plant failed to meet the 1972 Clean Water Act standards. The act required full secondary treatment, regardless of whether treated wastewater is discharged into a stream, river, lake, or ocean. The Point Loma plant uses enhanced primary treatment, thereby setting off decades of political and regulatory wrangling.

A large part of the problem is location. San Diego's wastewater-treatment plant is tucked into a carved-out section of a steep bluff, practically at the water's edge. Hemmed in on one side by the Navy and on the other by Cabrillo National Monument, upgrades to full

1

secondary treatment could cost more than $2 billion. Meanwhile, long-term monitoring and a study by the Scripps Institution of Oceanography have determined that discharges from the Point Loma plant have a negligible effect on the surrounding marine environment and have caused no fecal contamination of beaches. In contrast, untreated sewage and toxic chemicals from Tijuana contaminate San Diego's South Bay beaches on a regular basis. The irony is lost on no one, but not everyone is convinced by the comparison. Environmental groups have long disputed conclusions that discharges from the Point Loma plant are not harming coastal waters.

After passage of the Clean Water Act, San Diego and several other coastal cities sought modification of the act's one-size-fits-all approach. In 1977, Congress established a waiver process to provide some wiggle room. The burden was on the discharger to prove no harm, and waivers must be renewed every five years. The U.S. Environmental Protection Agency (EPA) tentatively approved San Diego's first waiver in 1981.

In 1986, the city's waiver was denied due to concerns about bacteria in kelp beds and effects on benthic marine life. The EPA, Sierra Club, and Surfrider Foundation sued the city, but treatment plant upgrades were again forestalled, this time by the passage of the 1994 Ocean Pollution Reduction Act. Despite its broad name, the act applies solely to San Diego, creating a special waiver process as an addendum to the Clean Water Act.

The new waiver came with a new mandate. By January 2010, San Diego was required to build enough capacity to reuse 45 million gallons per day of treated wastewater. The reuse would mean less wastewater dumped in the ocean.

At the same time that San Diego is dumping wastewater into the Pacific Ocean, the city is short on local water resources. The cause of this water scarcity problem is twofold—location and climate. San Diego sits in a cul-de-sac at the southwestern corner of the continental United States, bounded by the Pacific Ocean to the west, Mexico to the south, and mountains and desert to the east. Although it's not obvious from looking at San Diego today, the city has a semi-desert climate. Average annual rainfall is about ten inches, and some years the city receives less than half that amount. Most of the rainfall typically comes during just a few days of the so-called rainy season.

San Diego's challenges with water availability go back to the earliest days of European settlement. Father Junipero Serra established California's first mission in San Diego in 1769. Five years later, the mission was moved six miles east to ensure a year-round water supply from the San Diego River.

Seven stream systems that originate in the mountains of San Diego County and drain to the Pacific Ocean are the primary local source of freshwater. Beginning in the late 1800s, the county witnessed one of the nation's earliest and most prolific dam-building sprees. A few were record-breaking. When Sweetwater Dam was completed in 1888, it was the highest masonry arch dam in the United States.[3] *Harper's Weekly* reported that Morena Dam, completed in 1912, was the biggest dam in America, and possibly the world.[4] The reservoirs allowed the city to maintain an adequate water supply until World War II. As a center of aircraft production and a Navy town, it became clear that the city's local water resources no longer could supply the demands of its rapidly growing population.

The only available large freshwater source was the Colorado River, some 200 miles to the east. In addition to the distance, high mountains and a desert in between made it prohibitively expensive for San Diego to tap into the river on its own. Fortuitously, the Metropolitan Water District of Southern California (MWD) had recently completed the Colorado River Aqueduct into the Los Angeles region. San Diego could tap into MWD's system simply by building a connecting aqueduct up the coast. So that's what they did, but not without considerable handwringing over concerns that San Diego would end up perpetually playing second fiddle to Los Angeles for its water. The aqueduct came online in 1947—and none too soon. With less than three weeks' supply, San Diego was on the brink of running out of water.[5]

Thus began San Diego's dependence on MWD for most of its water supply, a relationship that's been described as a "shotgun marriage, where divorce is impossible."[6] The aqueduct was initially sized to meet only the Navy's projected water needs. It has required multiple expansions over time, including tapping into MWD's infrastructure to bring water from northern California and from Imperial Valley farmers.

This situation reached something of crisis proportions during the severe 1987–1991 drought, when the MWD board voted to cut deliveries to all customers in half. The proposed cut would have been devastating to San Diego, and the city (and county) felt betrayed by the proposed across-the-board cuts. With its reservoirs rapidly emptying and dependence on imported water for 95 percent of its supply, San Diego's need for water was much more critical than Los Angeles's. In the nick of time, a "miracle March" rainfall in 1991 came to the rescue, but the drought and MWD's response was a wake-up call. San Diego became determined to move away from its almost-monopolistic dependence on a single water supplier. "Never again! No more water shortages!" became the rallying cry.[7]

The bottom line is that San Diego is trying to kill two birds with one stone. By developing capacity for significant wastewater recycling, the city can avoid expensive upgrades to its wastewater-treatment plant *and* reduce its dependence on MWD. What's not to like?

San Diego is no neophyte to water recycling. In 1981, the city began testing the use of water hyacinths for secondary treatment—a novel approach adapted from NASA. The space agency had been exploring various plants to filter sewage and generate oxygen for future space stations. "The hyacinth's favorite place to grow is in raw sewage," noted Bill Wolverton, the principal researcher at NASA. "Throw it in a sewage lagoon that is cruddy and yucky, and it will thrive."[8]

As wastewater flowed slowly through long ponds filled with water hyacinths, these fast-growing aquatic plants naturally absorbed pollutants. Each plant dangled a thick mass of roots that resembled a bottle-cleaning brush. Thousands of tiny root hairs hosted bacteria that broke down dissolved organic matter. The hyacinths would then be harvested and used for fertilizer or burned for energy. "The idea was that you could use everything, with nothing left to dispose of," noted project director Paul Gagliardo.[9]

Hyacinths had several other advantages over conventional secondary treatment, including better removal of heavy metals and toxic organic chemicals. The approach also required little energy compared to conventional treatment, which requires energy for stirring and aeration of bacteria. (Today, treatment plants often meet their energy demands by generating their own renewable energy using methane digesters.)

The ponds provided recycled water for agricultural and highway irrigation. The city had hopes that, with additional filtration using reverse osmosis and other steps, the recycled water could someday be used for drinking water. But there were a few downsides. The ponds take up a lot of land—not something San Diego has in excess. The slow-moving, nutrient-rich water also bred huge numbers of mosquitos. Mosquito fish were introduced to eat the larvae, but the fish tended to die off.[10] Finally, the plants didn't grow well in winter, even in San Diego's mild climate. The city eventually dropped the idea.[11]

In 1991, San Diego began studying methods of advanced water purification, including reverse osmosis, granular activated carbon, and ultraviolet radiation. The reclaimed water met all drinking water standards. A study of the health effects found no evidence that the treated wastewater posed an elevated health risk to the general public. Before disinfection, the treated effluent was actually less likely to contain high concentrations of microbial indicators for infectious disease than the existing water supply.[12]

The studies did, however, reveal some cautionary signs. In theory, reverse osmosis (RO) should remove all viruses because these systems are designed to remove very small molecules. Early test results, however, showed that RO sometimes incompletely removed viruses seeded in the feed water.[13] This finding highlighted the importance of *redundancy*—use of multiple treatment barriers to attenuate the same type of contaminant—for example, use of disinfection in addition to RO to remove viruses. Redundancy has become a basic tenet of potable reuse projects. RO membranes also have improved over time.

The health effects study used short-term bioassays to evaluate genetic toxicity and potential cancer-causing effects. Organic extracts from reclaimed water and local reservoir water sources both exhibited some genotoxic activity. The activity, however, was stronger in the reservoir water than the reclaimed water. A moral of this story is that it's impossible to establish zero risk for any drinking-water source.[14]

In 1993, San Diego decided to move beyond the pilot phase, proposing the first project in California to use reservoir augmentation, instead of groundwater, as the environmental buffer for indirect potable reuse. Known as the Water Repurification Project, about half of the recycled wastewater used to meet the Point Loma waiver would undergo advanced wastewater treatment. This water would then be pumped twenty-three miles to a reservoir, where it would spend about a year blending with other water in the reservoir. Water withdrawn from the reservoir would undergo conventional drinking-water treatment before being distributed to customers. The project would be up and running by 2002.

San Diego utility managers understood the critical importance of public acceptance and embarked on a comprehensive research project to better understand the potential issues that needed to be addressed. Focus groups tested monikers for the water, ultimately choosing *repurified* over *recycled*. The latter apparently left too much to the imagination. Once it was fully explained to them, many participants supported using *repurified* water for drinking, washing, and cooking. All elected officials were notified about the project and told that the city and its consultants would meet with anyone who wanted to know more. Public outreach included brochures, video presentations about the project, feature stories in newspapers and other media outlets, and a telephone inquiry line.[15]

The Water Repurification Project seemed to have support from the public and community organizations. In August 1997, the *Los Angeles Times* reported that "San Diego leads the state" in water recycling.[16] "The only question appeared to be affordability. There were lots of supporters. It was all good," recalls Patricia Tennyson, a communications consultant who has been involved in reuse projects for decades.[17]

In March 1998, the California Department of Health Services certified the reuse program as "acceptable," but the tide was turning as the media began to sensationalize the idea of turning sewage water into drinking water. A prime opportunity arrived during "sweeps week," the time of year when Nielsen takes its survey of television viewing habits. With advertising rates on the line, producers do whatever they can to ratchet up their ratings during this period.

Marti Emerald, a television consumer affairs reporter in San Diego, saw an opportunity to entice viewers with a multi-part series on potable reuse. While visiting a project in the Los Angeles area, she heard the term *toilet to tap*.[18] For a prop, she used a toilet connected to a mysterious black box, then to a drinking faucet. Viewers were asked, "Do you have confidence that magic happens in this box?"[19] To make sure everyone got the message, an actress was shown spewing water out of her mouth in disgust.

The "toilet to tap" moniker was quickly adopted by public figures and reporters. Politicians running in upcoming state and local elections took advantage of this golden opportunity to take a stand against "toilet to tap" and attacked opponents who had gone on record supporting the Water Repurification Project.

Up for reelection, state assembly member Howard Wayne portrayed himself as the first politician willing to expose the "financial boondoggle."[20] He sent out a questionnaire asking his constituents if they supported drinking their own toilet water.[21] Not surprisingly, considering how the question was posed, about 80 percent of responses were negative. The questionnaire helped motivate about three hundred people to attend a hearing on the issue.[22] Wayne won his closely contested race.

Brian Bilbray, running for the House of Representatives, latched on to the toilet-to-tap issue to disparage his Democratic opponent's record. Bilbray, an avid surfer, went against the position of his surfer bros, most of whom supported the Water Repurification Project. Despite the environmental reputation of Bilbray's coastal district, however, voters were less likely to vote for a candidate who supported the reuse project. Bilbray squeaked by in the election, his opposition to potable reuse quite possibly making the difference.[23]

Bruce Henderson, running to recapture a city council seat that he formerly held, compared the repurification plan to a "Dr. Frankenstein" experiment without people's consent.[24] Henderson lost to the incumbent by a two-to-one margin.

Several politicians circulated the idea that the treated wastewater would be piped from affluent communities, like La Jolla, to economically depressed communities. The idea that poorer people would be exposed to health risks was widely covered by the media.[25] The environmental-justice accusations were untrue. The treated wastewater would be distributed to a wide range of communities, but the truth didn't begin to take hold until after the election.

In addition to politicians, Daniel Okun, a well-known environmental science professor at the University of North Carolina, became an outspoken voice of opposition. Okun, a member of both the National Academy of Engineering and the Institute of Medicine, lobbied for nonpotable reuse. He cautioned that "as with anything of this sort, we depend upon technology to protect us, and there are failures of technology."[26]

Okun raised concerns about unknown contaminants that might pass through the treatment barriers. Among the more troubling were endocrine disrupters—chemicals that interfere with the actions of hormonal systems. As the debates about San Diego's potable reuse project were ongoing, researchers were reporting widespread findings of intersex fish showing both male and female characteristics.[27] There was no evidence that exposure to these chemicals at trace levels in drinking water was a human health problem, but it was a growing topic of public concern at the time and one that easily caught people's imagination.

In early 1996, control of the project was transferred to the city's wastewater department, which was thought to be better able to fund and construct the project. The wastewater staff interacted with members of the environmental community but had little regular interaction with the broader community and were poorly suited to promote the project to San Diego's residents. This reinforced the perception that the potable reuse project was more of a convenient scheme for addressing wastewater disposal than a strategy for ensuring a safe and reliable drinking-water supply.[28]

There was also a competing project for increasing San Diego's water supply. In 1998, the San Diego County Water Authority (the agency that delivers water to the City of San Diego and other member agencies) was negotiating a long-term agreement with Imperial Valley farmers. If successful, it would become the largest rural-to-urban water transfer in U.S. history. Farmers in the Imperial Irrigation District (who hold almost three-quarters of California's water rights to the Colorado River) would be compensated for implementing agricultural water-conservation measures, as well as lining two major irrigation canals to reduce leakage. In return, the freed-up water would be transferred to San Diego using MWD's aqueducts. The San Diego County Water Authority saw this as the centerpiece of its future water supply. Water reuse was a much lower priority. A return to wetter weather during this period also didn't help, as memory of the drought that helped motivate the reuse project quickly faded.

And then came the Revolting Grandmas, a particularly outspoken group opposing the potable reuse program. The "group" was basically comprised of two individuals: Muriel Watson, a retired schoolteacher, and her friend Mary Quartiano. In early 1998, they decided to "educate" the public and lawmakers on how, in their view, to properly recycle sewage.

Watson had a flair for the dramatic. An ardent Republican, she ran unsuccessfully four times for state senator. She also founded *Light Up the Border* in 1989, urging citizens to drive their cars to the U.S.–Mexico border and shine their headlights on illegal border crossers. This initiative soon peaked at over two thousand cars and resulted in stadium-style lights being installed along a two-mile stretch of previously unlit border fence.[29]

The Grandmas adamantly opposed potable reuse and quickly adopted the toilet-to-tap moniker. They lobbied for nonpotable uses such as landscape irrigation—or "showers-to-flowers" as they dubbed it. They argued that the money could be spent much more wisely on purple pipes and desal plants, and let's not forget fixing potholes! They were media savvy, sending out "news releases" attacking the potable reuse plan and calling on citizens to participate in city council meetings. "Everyone who does not care to drink toilet water should be there," they urged.[30]

Finally, there were the dueling advisory panels.

In 1998, the National Research Council completed a nationwide report on potable reuse. The general conclusion was that indirect potable reuse is viable when there is a project-specific assessment of the sort that San Diego had completed, but also cautioned in the executive summary that "indirect potable reuse is an option of last resort."[31] It should be adopted only if other measures have been evaluated and rejected as technically or economically infeasible. The statement "option of last resort" became the major finding of the report in public discourse.

In 1998, two other science panels reviewed the San Diego project. One consisted of national experts and the other of local scientists assembled by the San Diego County Board of Supervisors. The national panel recommended moving forward with the project. The local panel did not, due to concerns about the ability to adequately remove endocrine disrupters and other contaminants of emerging concern in wastewater.

The project and findings of these two panels were reviewed by the San Diego grand jury—a panel of citizens appointed annually to serve as an oversight committee on county government activities. The grand jury concluded that the concerns of the local science advisory panel about emerging contaminants were merely hypothetical. They had never been witnessed where indirect potable reuse had been practiced. The grand jury report was clear in its view: "It is time for the City Council to take a position of leadership and to make policy which will result in the development of additional sources of water. The need of future generations should outweigh personal and public opinions as well as political expediency."[32]

The recommendation fell on deaf ears.

In December 1998, Mayor Susan Golding froze any further discussion of potable reuse. Golding cited concerns about safety and costs. San Diego had spent more than $15 million, accumulated two decades of scientific research, and waged a public relations campaign to convince politicians as well as the general populace that wastewater can be made safe to drink. "But all that might be going down the drain," reported the *San Diego Union-Tribune*.[33] The following month, the city council directed the city manager to put the project on indefinite hold.[34] It became taboo for city officials to even discuss it.

Golding's announcement came as a surprise to almost everyone. The program manager learned about it the next day while he was reading the morning newspaper. But political consultants saw it as inevitable. "Toilet to tap is hazardous to the political health of anyone who touches it," claimed one political consultant. "It's one of those deals where the people who support it don't really care about it, while people who oppose it care a great deal—which is what makes it a good campaign issue," said another.[35]

As far as opponents were concerned, they had won. Toilet to tap had crashed and burned. In reality, potable reuse had gone underground and would rise again.

Chapter Two

Early Days

All the way along, we got people pushing us to let them use the lakes, rather than our trying to persuade them to come in.

—Ray Stoyer[1]

When Jimmy Lowery jumped into that swimming pool one summer's day in 1965, the splash he made reverberated well beyond the pool. The nine-year-old boy became the first person in the United States to swim in reclaimed sewage water. While Jimmy's splash was remarkable, even more extraordinary is that not only did recreational use of treated wastewater have full public acceptance, but also the people of Santee, California, even clamored for it. One man's vision and dogged pursuit made it all possible.[2]

In the 1950s, Santee, a small town fourteen miles northeast of San Diego, was rapidly converting from farmlands to suburban housing developments. The town had a relatively new treatment plant that discharged secondary-treated wastewater into Sycamore Creek, a tributary of the San Diego River. When the state tightened the standards for Santee's wastewater discharges in 1959, the town faced two choices: Make expensive (and at the time technologically risky) upgrades to its sewage-treatment plant, or tie into San Diego's Metropolitan Sewer System. The latter option required signing a forty-year open-ended contract with San Diego. Neither option appealed to the town.

Santee also faced water-supply challenges. Construction of El Capitan Dam in the 1930s had captured much of the flows of the San Diego River that previously recharged the area's alluvial aquifer. As a result, Santee was completely dependent on imported water purchased from the Metropolitan Water District of Southern California. Given that imported water was only going to get more expensive, Ray L. Stoyer, manager of the Santee County Water District, thought, Why not reuse the water they already had? Stoyer was willing to gamble on a natural way to further clean Santee's wastewater for reuse.

Stoyer took frequent walks up the valley from the treatment plant. It was hardly an idyllic setting. Mining of sand and gravel for the local construction boom had left behind piles of soil covered with weeds and surrounded by large unsightly pools of stagnant water. While walking one day, Stoyer suddenly envisioned transforming this unsightly mess into a chain of lakes. The town's sewage-treatment plant would provide the water, and the lakes would provide natural oxidation. This relatively simple plan would avoid the need for expensive upgrades to Santee's treatment plant or chaining themselves to San Diego's sewage system.[3]

Stoyer approached the owner, Bill Mast, about donating the excavated sand and gravel pits to the town. There was something in it for Mast, who owned a parcel of land at the mouth of the canyon that could be turned into a golf course irrigated by the lake water. Mast not only agreed to donate the sand and gravel pits but also contributed to the eventual aesthetics by leaving mounds of sycamore and oak trees as shady islands or peninsulas.

Once Stoyer convinced his board members of the idea, it was up to him to make good on the gamble. Three large lakes were created just north of the treatment plant. Sewage plant effluent was pumped into the first lake, sat for a period, then was pumped to the second lake and so forth. Coliform tests confirmed that the water became purer as it moved up the valley from lake to lake. A fourth lake was soon added.

To get the public onboard with his vision of using the lakes for boating, fishing, and other recreational activities, Stoyer decided to tempt them with an attractive recreational spot. Grass and shrubs were planted on the banks of one of the lakes, and picnic tables were set around its shores. The lake was stocked with sunfish and bass. Stoyer even imported a few ducks. A high wire fence allowed people to see but not use this enticing waterpark in the arid valley.

When people asked why they could not enter the lakes, Stoyer explained that the district would not even consider opening it up until health authorities ruled that the water was absolutely safe for human contact. This helped put people at ease about someday using the lakes. Stoyer also embarked on an extensive speaking and publicity campaign, giving talks to local groups and interviews with local papers. In the meantime, homes bordering the lakes were selling well. According to author Leonard Stevens, this was "undoubtedly the first time in history that a sewage treatment system drove up the real estate prices in its immediate vicinity."[4]

Stoyer's biggest challenge was to convince the county health authorities about the project's safety. Dr. J. B. Askew, the health department director, flat-out rejected the proposal, saying: "You cannot let children around a body of water, before they are in it. At least their hands are in it, and the next minute their hands are in their mouths."[5]

Not one to quit, Stoyer added an extra natural treatment step. The water district pumped the water a mile up the creek to percolation beds that they created at an unmined site. From there, the water flowed downhill through sand and gravel to a ditch that channeled it to the uppermost lake. After tests demonstrated the effectiveness of the percolation beds, Askew agreed to allow boating but insisted that people stay out of the water. Stoyer's success at creating pent-up demand became immediately obvious at the grand opening in 1961, when thousands of people flocked to the lakes for picnicking and boating. Recreational fishing was allowed the following summer, although people were forbidden to eat the fish for several years.

Before anyone was allowed to enter the water, scientists had to demonstrate that it was safe from viruses. Virologists undertook a three-year study with the support of federal and state health authorities. By coincidence, Santee residents were taking the oral Sabin polio vaccine as part of a nationwide vaccination campaign. After receiving the vaccine, a person's feces contained large numbers of polio virus, providing an excellent test of the ability of the system to remove them. The inactivated virus also was injected in water at the high end of the percolation bed. The study concluded that the risk from waterborne viruses was extremely low. With these results in hand, water was pumped from one of the lakes to a newly constructed swimming pool, where the water quality could be closely monitored. Jimmy Lowery may have taken the first plunge, but the pool was quickly packed with happy, splashing kids.[6]

Currently, the Ray Stoyer Water Reclamation Facility recycles about two million gallons of water each day.[7] The treated effluent flows through seven interconnected lakes, creating a unique recreation area with over 760,000 visitors annually. Although the swimming is long gone, amenities include camping, cabin rentals, fishing, boating, playgrounds, and walking trails. About half of the reclaimed water is used for irrigating city parks, schools, and the Carlton Oaks Country Club (the golf course formerly owned by Mast). Over the years, delegations from around the world have come to see and learn about Santee Lakes.[8]

The Santee water district did not completely free itself from San Diego's Point Loma wastewater-treatment plant. By the 1970s, the lakes and nonpotable reuse could no longer accommodate all of Santee's reclaimed sewage. Excess discharge flowed into the San Diego River via Sycamore Creek, causing dense bush growth and interfering with mosquito abatement. In 1974, the state mandated that Santee's excess effluent had to be pumped to the San Diego Metropolitan Water District—the same agency that prompted Santee officials to develop the lakes.[9]

In 1976, the Santee water district merged with another water district to form the Padre Dam Municipal Water District. Padre Dam continued to build on its recycling legacy and has become a leader in potable reuse. We will return to this story in a later chapter.

Farther north, Irvine Ranch Water District (IRWD) in Orange County, California, is another pioneer in water reuse. Soon after it was established in 1961, IRWD made the decision to provide recycled water for nonpotable uses. The district began by delivering tertiary treated water to agricultural users. As master-planned suburban communities grew out of the former large ranch, IRWD installed water lines for nonpotable uses alongside water mains. This forward thinking avoided the expense and disruption of later having to retrofit built-up areas.

In 1991, IRWD earned California's first unrestricted use permit, allowing the recycled water to be used for almost anything, except drinking. The nation's first health department–approved building using reclaimed water for interior uses came online that same year.[10] Today, recycled water makes up more than a quarter of IRWD's total water supply. A small portion is still used for agriculture. The largest use is for landscape irrigation virtually anywhere that anything grows on public or commercial property. Parks, schoolgrounds, golf courses, freeway medians, common areas managed by homeowner associations, and some large residential lots all use recycled water. IRWD also serves recycled water to more than 130 dual-plumbed buildings where it's used to flush toilets and urinals. These range from a single restroom at a park to twenty-story high-rise office buildings and include the first hotel in the country where the toilet in each individual guest room uses recycled water.[11]

Among IRWD's claims to fame is the pioneering use of purple pipe. In the early 1980s, water district engineers were looking for a way to clearly distinguish the pipes that carry recycled water from those that serve drinking water. Initially, colored tape was placed along recycled water pipes. A local pipe manufacturer proposed colored pipe as a more reliable and simple approach. Keith Lewinger, assistant director of planning, was assigned the task to select among colors not already used for other designations. Lewinger is red-green color blind. He picked out the lavender color, because he could distinguish this shade of purple from other colors. If he could identify it, went the reasoning, so could everyone else. IRWD worked with the American Water Works Association to establish "Irvine Purple" as a national standard for recycled water.[12]

Today, IRWD recycles almost ten billion gallons of water each year. To help ensure that recycled water is available when needed, the district relies on four large seasonal storage reservoirs. But managing recycled water is a balancing act. During the cool, wet winter months, the community does a lot less watering, the reservoirs fill up, and the district can run out of room to store recycled water. Then, in the summer, when communities are watering again, the district at times doesn't have enough recycled water and needs to buy more expensive imported water to make up the difference. IRWD is proposing to expand one of its reservoirs (Syphon Reservoir) to allow the district to store all of the recycled water that it produces. Salinity also can present challenges, but the levels found in IRWD recycled water are lower than most, because fresh groundwater is the source of much of the district's potable water supply.[13]

While Irvine brought us purple pipes, it was not the first to use dual-distribution systems for nonpotable uses. That distinction belongs to a more awe-inspiring location.

Grand Canyon Village, on the South Rim of the Grand Canyon, attracts millions of visitors each year with its spectacular canyon overlook. Meeting the water needs of visitors to this semiarid high plateau has long been a challenge. The first settlers hauled water by burros, and later by tankers over roads and rails. An attempt to tap groundwater proved futile, with the hole still dry after drilling one thousand feet.[14] Drawing from the sediment-laden Colorado River far below the rim posed its own formidable challenges.

With every drop precious, a sewage-treatment plant was built in 1926 to provide reclaimed water for power generation, steam locomotives, and flushing toilets—giving the village the distinction of using the first dual-distribution system in the United States.[15] Engineers responsible for the system were concerned about the possibility of accidental cross-connections of the reuse water with a sink or water fountain, so they identified the pipes carrying nonpotable water with red paint (this was before Irvine Purple came into vogue). Dye was added to the reuse water once a year as an extra safety measure.[16]

The local power station and steam locomotives are long gone, but reclaimed water continues to be used for landscape irrigation and flushing toilets.[17] This recycled water supplements potable water that is piped twelve miles from a North Rim spring through an aging pipeline in the continuing challenges to supply freshwater.[18]

One of the first major potable reuse projects in the United States occurred not in the west, but rather in the east. Water quality, not water scarcity, was the motivating force.

The Occoquan Reservoir is the source of drinking water for more than a million people in the northern Virginia suburbs of Washington, DC. During the 1960s, the once-rural watershed underwent rapid suburban development. By 1970, eleven secondary-treatment plants were discharging inadequately treated sewage into tributaries of the Occoquan watershed. The reservoir was rapidly deteriorating with algal blooms, fish kills, and taste and odor problems.[19]

In a typical year, the effluent accounted for less than 10 percent of the overall flow—not atypical of many other cities. During dry periods, treated wastewater accounted for more than 80 percent of the water entering the reservoir. With residence times less than a year, people were drinking mostly secondary-treated wastewater during times of drought.

In 1971, the Upper Occoquan Sewage Authority (*Sewage* was later changed to *Service*) was created to provide state-of-the-art treatment for all wastewater generated in the watershed. The wastewater was rerouted to a single treatment plant, where it underwent "a water engineer's dream," observes University of California professor David Sedlak, who has written extensively on urban water and wastewater systems. "The plant's designers threw everything they could come up with at the wastewater: activated sludge, filtration, activated carbon treatment, ion exchange, chlorination, and lime clarification." After this expanded treatment, the water in the Occoquan Reservoir was probably better than water flowing in rivers and reservoirs downstream of many cities.[20] A 2001 independent study comparing pathogens in the reclaimed water with the reservoir water showed that, in every case, the treated wastewater was of better quality than the water it was mixing with in the reservoir.[21]

The plant began operations in 1978 at a relatively modest ten million gallons a day, expanding over the years with population growth to fifty-four million gallons a day. The Occoquan system is the oldest potable reuse project in the United States using reservoir augmentation. The reservoir water quality improved dramatically after the treatment came online. Today the reservoir is a healthy fishery and recreational area. A monitoring group that is independent of both the wastewater and drinking-water treatment plants has watched over the water quality during more than forty years of successful operations. Virtually no opposition has arisen, in large part because the advanced water-treatment plant has visibly improved the reservoir water quality.

The earliest reuse of wastewater (untreated) is for agriculture, with the ancient Greeks and Romans among the first practitioners. In the late nineteenth century, sewer farms sprang up in Australia and several European countries. These "farms" were primarily disposal operations with incidental use of the water for crop production.[22]

In the early twentieth century, southern California cities were also turning to sewer farms. By 1910, about three dozen California communities were using untreated sewage for irrigation, simultaneously profiting from human waste while sending it away from homes. In one of the first profitable sewer farms, Pasadena, California, purchased a 300-acre plot of land outside the city and piped in raw sewage to irrigate walnuts, pumpkins, hay, and corn.[23]

Although sewer farms in the United States soon disappeared, untreated wastewater is still applied to crops around the world, particularly in China and Mexico.[24] The world's oldest and largest use of untreated wastewater for irrigating food crops is in the Mezquital Valley north of Mexico City—a good reason not to eat uncooked vegetables when visiting Mexico City.[25]

In 1918, California published its first regulations for recycled water, prohibiting the use of raw sewage and septic tank effluents for irrigating food crops eaten raw.[26] Sporadic uses of treated wastewater for agriculture occurred in the following decades, but it wasn't until the late 1970s that a seminal study of the safety of using treated wastewater for irrigating food crops took place in Monterey County, California—home to the Salinas Valley. Known as the Salad Bowl of the World, extensive groundwater withdrawals for agriculture in the valley were causing large-scale saltwater intrusion. Community leaders began entertaining the idea of using treated wastewater for agricultural irrigation to slow the saltwater advance.

Environmental health officials and the farming community needed to be convinced of the safety of this approach to both consumers and farm workers. Local farmers also feared customer backlash against produce irrigated with "sewer water." They insisted that the produce not be labeled as having been irrigated with recycled wastewater.[27]

These safety concerns led to the Monterey Wastewater Reclamation Study for Agriculture. Bahman Sheikh, an Iranian immigrant who came to California as a graduate student, led this landmark effort. From inception of the study in 1976 through publication of its final report in 1987, Sheikh and his team investigated the fate and transport of pathogens from using disinfected tertiary-treated wastewater on food crops eaten raw. Sheikh also worked closely with local farmers and health authorities to communicate the study significance and garner acceptance of water reuse. He continued to be widely recognized as an international leader in the use of recycled water for irrigation until his death in 2020.

No naturally occurring virus was ever detected in any of the monthly samples of irrigation waters. To further verify the safety of water reuse, the investigators seeded the influent water to the pilot plant with vaccine-strain poliovirus (a deactivated form that poses no threat to human health). The results confirmed the goal of at least 99.999 percent reduction in viruses by the treatment process. The study also demonstrated that use of the treated wastewater did not lead to accumulation of metals in soils or plant tissue. The overall marketability, quality, and yield of crops was comparable with produce grown with other sources of irrigation water.[28] In 1998, the world's largest reclaimed water facility for irrigation of food crops eaten raw began supplying water to twelve thousand acres of prime farmland in the Salinas Valley.

The above examples provide snapshots of some of the early adopters of water reuse. Before turning to other examples in more detail, it's worth noting how the relationship of wastewater agencies with their communities has evolved since these early days. Camden, New Jersey, is a good example.

Located across the Delaware River from Philadelphia, Camden is one of the nation's poorest cities, with a history of pollution and environmental injustice. Prior to 1987, forty-five million gallons of inadequately treated sewage were discharged daily into the lakes and streams of Camden County. Today, the Camden County Municipal Utilities Authority (CCMUA) treats the sewage discharged from properties throughout the county.[29]

The CCMUA's former executive director, Andrew Kricun, is passionate about the utility's opportunities *and* obligation to serve the community. In the past, he says, the "ceiling of aspirations" for wastewater utilities was compliance with the EPA discharge permit. If you achieved that standard, you were good to go. Minimal interaction with the community was the norm. "I thought we should graduate from doing no harm to being a proactive good neighbor. Then, graduate from being a good neighbor to ultimately striving to be an anchor institution in the community," says Kricun.[30]

He and the CCMUA walked the talk. Among its community services, the CCMUA has partnered with Americorps to provide jobs to at-risk youth to help maintain the city's green infrastructure of restored streams and rain gardens that capture runoff. Up to sixty at-risk young men and women benefit from the program each year, which includes life-skills training as a key part of the package. The CCMUA also provides green summer jobs for ten to twenty Camden high school students with an interest in environmental science.[31]

Wastewater-treatment plants are notoriously energy intensive, with up to 40 percent of customers' sewer bills going toward energy costs.

The CCMUA is working toward energy self-sufficiency. The agency first reduced energy usage by approximately 20 percent through investments to improve the plant's efficiency. Installation of a solar panel array contributed another 10 percent. A third contribution comes from extracting biogas from the plant's sludge and then running the gas through an electric turbine.

The final step toward energy independence involves an innovative water-reuse project. Treated effluent will be piped to a nearby waste-to-energy incinerator for use as cooling water. In return, the incinerator will send electricity generated from burning trash to the CCMUA treatment plant. It's a win-win. The CCMUA needs energy to generate clean water, and the incinerator needs water for its steam turbines to generate energy. The treated effluent from the CCMUA plant will replace a million gallons per day of groundwater that is withdrawn from an overstressed regional aquifer now serving as the county's primary source of drinking water.

Through low-interest loans from the state, the utility has been able to undertake its green energy improvements without raising rates to its customers. The CCMUA has won several awards for its efforts and is recognized by the EPA as a "Net Zero Hero."

Kricun likes to quote the nineteenth-century English novelist George Eliot in *Middlemarch*: "The growing good of the world is partly dependent on unhistoric acts [of faithful men and women]." The Eliot quote is appropriate on multiple levels. George Eliot was the pen name used by Mary Anne Evans to ensure that her works were taken seriously. In the mid-1800s, women writers were considered just romance novelists and went unrecognized for their important societal contributions. Many public servants who work for water utilities today contribute more to society than is recognized by the public. The water-recycling revolution is, in large measure, the result of many underrecognized individuals.

Chapter Three

Trailblazers

To meet all needs—domestic, agricultural, industrial, recreational—
we shall have to use and reuse the same water, maintaining quality
as well as quantity.

—President John F. Kennedy, 1961 Speech to Congress[1]

The Los Angeles area's groundbreaking work in water recycling is
a striking example of geology bumping up against human ingenuity.
Much of southern Los Angeles County sits on a large coastal lowland
known as the LA Basin. To the northeast lies the San Gabriel Valley,
separated from the LA Basin by a ridge of hills. The valley's two main
rivers, the Rio Hondo and the San Gabriel River, flow into the LA Basin
through a two-mile-wide gap in the ridge known as the Whittier Narrows. Think of it as a funnel with the Narrows as the spout.

Los Angeles relies on groundwater for much of its public water supply, but the geology works against them in a critical way. Aquifers in
much of the LA Basin are overlain by fine-grained sediments that allow
little natural recharge to the groundwater system. When Los Angeles
began its explosive growth after World War II, it didn't take long before
groundwater levels began dropping precipitously.

There is, however, a place where the geology works in their favor.
Just below the Whittier Narrows, in an area known as the Montebello
Forebay, are unconfined aquifers with deep sandy sediments. The

subsurface is permeable all the way down to the water table, making the area well suited for recharging the groundwater system.

Los Angeles solved their impending water crisis by siting spreading basins along the Rio Hondo and the San Gabriel River in the Montebello Forebay.[2] Diversion of stormwater to the spreading basins began in the late 1930s.[3] Upstream dams designed for flood control helped capture floodwaters that would otherwise be lost to the ocean. In the 1950s, imported water from the Colorado River was introduced to help recharge the spreading basins. In 1962, recycled wastewater was added to the mix. There are no pumping costs because the rivers act as a conveyance from upstream wastewater-treatment plants to the spreading basins. Thanks to this made-to-order combination of geology and human ingenuity, groundwater levels in the LA Basin began to recover, and Los Angeles dodged a major water-supply crisis.

The Montebello Forebay has the distinction of being the oldest planned potable reuse operation in the United States—although there is a possible competitor. In 1924, the City of Fresno, California, installed wells for the purpose of lowering groundwater levels to increase percolation capacity for wastewater.[4] Because groundwater was the primary drinking-water supply for the city, this project arguably created the state's first planned potable reuse system.[5]

In 1989, after almost three decades of successfully treating, capturing, and recharging wastewater into the LA Basin's groundwater supply, Earle Hartling had an idea. Hartling was (and still is) the water-recycling manager of the Sanitation Districts of Los Angeles County. The County Sanitation Districts (for short) is a regional agency that serves over half the residents of Los Angeles County. It operates ten "upstream" water-reclamation plants and one facility that discharges secondary-treated wastewater to the ocean. Hartling's idea was to build a pipeline from one of the reclamation plants up into the San Gabriel Valley, so that those communities also could benefit from recycled water.[6] The San Gabriel Valley was experiencing rapid growth, and there was already a suitable location being used for recharging stormwater and imported water. All they needed to do was run a nine-mile pipeline. Simplicity itself.

Until things suddenly got complicated.

Officials at a Miller Brewing Company plant that drew its water from a nearby wellfield claimed that the project would "irreversibly pollute" the basin and contaminate their water supply.[7] In reality, only about 2 percent of the groundwater that Miller pumped would be of recycled water origin. The water was tertiary treated to the highest level then approved by the California State Health Department and would be further purified by filtration through the subsurface. Similar recharge for three decades below Whittier Narrows had reported no problems, health or otherwise. Miller also used advanced techniques to treat their water prior to their brewing process. Although reluctant to admit it in public, Miller's primary concern was loss of market share from bad press.

Miller filed a lawsuit to stop the project, and their public relations team kicked off a campaign to discredit the idea. According to Miller, the recharge plan could increase the breeding ground of disease-carrying mosquitoes. It could threaten sensitive plant and animal life. It could cause health problems, especially for infants.[8] The health ploy was pretty rich, coming from an alcoholic beverage producer whose parent company was Philip Morris, one of the country's largest tobacco companies.

The grand slam was three little words that continue to plague potable reuse efforts to this day—*toilet to tap*. The term originated from the opposition to the San Gabriel project, although there's some debate about who first used the phrase. The term is usually attributed to the Miller public relations staff.[9] Forest S. Tennant, a local pain doctor and former mayor of West Covina, also claims credit.[10] Tennant used his connections and money to create a nonprofit citizens group, "Citizens for Clean Water." In late 1993, the group took out several full-page newspaper ads against the project, proclaiming that it was unnecessary and potentially could cause cancer, dementia, and other diseases. The phrase "toilet to tap" did not appear in these ads.[11] The first mention of "toilet to tap" in print came in an in-depth article in the *LA Times* on the proposed San Gabriel Valley project and the objections raised by Miller Brewery and other opponents.[12]

Earle Hartling went from one public meeting to another trying to undo the damage. If the efforts of Miller Brewing and Citizens for Clean Water weren't problematic enough, he was confronted by a fellow named E. T. Snell, who showed up dressed as a clown, complete with frizzy green wig and white makeup. At one meeting, Snell took the podium and proclaimed, "This is a plot by the Trilateral Commission secret world government to poison the San Gabriel Valley." Then he pointed at Hartling and said, "If this water is so good, why doesn't this guy drink it?" In Hartling's hands was a bottle of sparkling-clear recycled water. Without missing a beat, he opened the bottle and drank it.[13] Snell continued to follow Hartling from meeting to meeting, lambasting both the project and Hartling personally. In later years, he was arrested and booked on multiple felony accounts for assaults on officials.[14]

The brewery jumped into political races for the board of the Upper San Gabriel Valley Municipal Water District, the recycled water project sponsor, contributing money to candidates who opposed the project. They focused negative press on Anthony Fellow, an environmentally oriented advocate of water recycling who was running for reelection on the water district's board. Miller sent out flyers with Fellow's face on a toilet and took out ads with sensational language.[15]

Ironically, given that the whole thing was about avoiding bad press, Miller made such a big deal that they kicked off a feeding frenzy of bad press. Like all brilliant bumper-sticker messages, the *toilet-to-tap* denouncement resulted in the media and public alike sitting up and taking notice. The jokes began about Miller beer being "aged in porcelain," as opposed to "beechwood aged."[16] Miller's ultimate nightmare occurred when Jay Leno made the toilet-to-tap story the butt of jokes on national television for two weeks running. Two and half years later, a compromise with Miller was worked out and even supported by Tennant, but given the bad publicity, the water district gave up on the idea.[17] Meanwhile, Anthony Fellow defeated his opponents three to one and remains on the water board to this day.

About the same time that the San Gabriel Valley recycling plan was going down, the City of Los Angeles was planning its own water recycling effort in the San Fernando Valley. This project would not only reduce the need for imported water but also help address a landmark decision at Mono Lake. It's an interesting case study of the intersection of environmental law, the public trust doctrine, and water reuse.

Mono Lake sits at the eastern foot of the Sierra Nevada to the north of Owens Valley. Five streams from the Sierra Nevada flow into the lake, but none leave. Having no outlet, Mono Lake is saline from the concentration of salts by evaporation. It's also part of an extraordinary ecosystem. Brine shrimp unique to Mono Lake number in the trillions during their late summer peak. The shrimp and alkali flies that ring the lake's edge provide food for nearly a hundred bird species, making Mono Lake a key stopover for migratory waterfowl along the Pacific Flyway. Islands in Mono Lake provide a breeding sanctuary for about a quarter of the California gull population.

In 1941, Los Angeles began to direct about half the flow of four of the lake's tributary streams into its Owens Valley aqueduct. A second tunnel in 1970 led to diversion of almost all the flow of the four streams. The diversions resulted in widespread negative impacts in and around the lake. From 1940 to 1980, the volume of water in Mono Lake shrank by 50 percent, and its level dropped forty feet.

In 1978, an ornithologist named David Gaines founded the Mono Lake Committee with a small group of students and others who were passionate about the lake. The Mono Lake Committee joined the National Audubon Society and other allies to file suit against the Los Angeles Department of Water and Power using a novel legal tactic. They claimed that the diversions from the Mono Lake basin violated the doctrine of "public trust."

The public trust doctrine is rooted in the idea that certain natural resources—such as the ocean and some water bodies—belong to the public. Because of their immense importance to individuals and society as a whole, no private entity should monopolize or deprive the public of the right to use and enjoy them. The government serves as the trustee of these resources for the benefit of the people.[10]

In 1971, the California courts recognized that the public-trust doctrine protects the ecological as well as recreational and economic values of tidelands and navigable waterways. The landmark Mono Lake case sought to expand the doctrine to nonnavigable waters—the tributary streams to Mono Lake—whose diversions harmed a navigable water body, namely, Mono Lake.

The environmentalists undertook an extensive public education campaign, including thousands of bumper stickers with slogans like "Long Live Mono Lake" and "Mono Lake: It's for the Birds." Stories about Mono Lake appeared in *Harper's*, *National Geographic*, *Time*, *Smithsonian*, *Audubon*, and *Sports Illustrated* magazines. The committee grew to more than twenty thousand members.

In 1983, the California Supreme Court ruled in favor of the environmentalists, concluding that the state has a dual mandate to balance the need for municipal water supplies with the need for water to restore and maintain natural water-dependent ecosystems. The existing prior appropriation rights of Los Angeles had been established without consideration of the environmental consequences and needed to be reevaluated.

In 1994, sixteen years after the Mono Lake Committee had initiated the struggle, the Los Angeles Department of Water and Power (DWP) and the Mono Lake Committee finally reached an agreement. The DWP agreed to a substantial reduction in diversions. A key part of the agreement was that it would not trade the health of Mono Lake for that of another ecosystem. Some of the replacement water for the reduced Mono Lake diversions would come from conservation.[19] Much of it would come from the planned East Valley Water Recycling Project in the San Fernando Valley. Everyone involved was very much on board with this idea—the Mono Lake Committee, the DWP, the Sierra Club, and the Los Angeles Chamber of Commerce, among others. In jubilation, Martha Davis, the executive director of the Mono Lake Committee, called the agreement "the political equivalent of the Camp David accord."[20]

The East Valley Water Recycling Project was designed to pump treated wastewater about ten miles from the Tillman Water Reclamation Plant in Van Nuys (one of four wastewater reclamation plants serving the City of Los Angeles) to spreading basins in the east San Fernando Valley. The tertiary-treated water would be further cleansed naturally during its five-year journey of more than a mile to the water agency's wells. The pumped water would then be chlorinated, mixed with water from other sources, and piped to customers.[21]

At first, the project moved along smoothly. In 1991, when public hearings were held on the Environmental Impact Report, and again a few years later, only a couple dozen people showed up. Aside from basic issues, such as traffic impacts during construction and concerns about promoting development in the valley, there was virtually no opposition—certainly none that was organized. The Los Angeles City Council approved it unanimously in 1995, and the *Los Angeles Times* ran a full-page article clearly describing the project.[22] Bill Van Wagoner, the engineer in charge of the program, says that during this time he received no calls of concern from the public.[23] Despite the toilet-to-tap rallying call in the neighboring San Gabriel Valley and later in San Diego, the project remained uncontroversial.

Things changed dramatically in 2000, just as construction was completed and the project was about to start. The public uproar began when the *Los Angeles Daily News* published an article with the headline "Tapping Toilet Water." The article was instigated by Gerald Silver, president of Homeowners of Encino. The Encino association aggressively fought traffic, sign blight, overdevelopment, and any other issue that affected "the single-family habitability" of this well-off community. Silver was concerned that the project might open up more development in the valley and claimed it would poison the water supply of the San Fernando Valley. He revived the toilet-to-tap mantra from the San Gabriel Valley controversy.[24]

Politicians quickly became involved. Los Angeles mayor Richard Riordan tried to distance himself from the project by saying he didn't recall it—until he was reminded that he and Governor Pete Wilson had promoted the plan at a 1994 news conference.[25] The issue became embroiled in the 2001 Los Angeles mayoral contest and a ballot measure calling for secession of the San Fernando Valley from Los Angeles. Mayoral candidate Joel Wachs, who had approved the plan in 1995, compared reclaimed water to aerial spraying of malathion.[26] Antonio Villaraigosa, another mayoral candidate, also denounced the project.

Similar to San Diego, opponents claimed that the east San Fernando Valley was a dumping ground for Los Angeles wastewater generated by wealthier communities. State senator Richard Alarcon, who had approved the project when he was a city council member, held a hearing on environmental justice concerns. Wachs admonished: "Go tell somebody in North Hollywood that they have to drink toilet water, but the mayor won't have to drink it in Brentwood" (an upscale neighborhood on the west side where the mayor lived).[27] Secession advocates claimed that the east valley would drink toilet water while the west side would get the "good" water. In reality, the spreading basins existed on the east side because of the favorable soils that allow percolation, and the water would be distributed to homes over a wide area.

City attorney James Hahn, planning his own mayoral run, ordered the project shuttered for no apparent reason beyond the public protest. Upon winning the mayoral race, Hahn made the shutdown permanent. Built at a cost of $55 million, it had been used for just a few days prior to being shut down. "We spent slightly under $1 million per acre-foot of water produced before we had to shut it off," recalls Van Wagoner.[28]

That comes down to about $3 a gallon—as opposed to the fraction of a cent per gallon usually paid by DWP customers. For decades, the severed pipeline at the spreading ground remained as a monument to the failed project.

But all has not been lost. The city resurrected the East Valley project in 2007 as California was in the midst of a major drought. Having defeated James Hahn for Los Angeles mayor in 2005, Antonio Villaraigosa reversed course and was now a leading advocate for purifying wastewater and returning it to the drinking water supply. After more than a two-decade setback, the initial phase of groundwater replenishment using the same pipeline and with ozone added to the treatment train is scheduled for 2021.[29]

A third proposed water-recycling project in the 1990s in the LA Basin would turn out quite differently. In response to the severe 1987–1992 drought, the West Basin Municipal Water District received state and federal funding to build a world-class water-recycling facility near the Los Angeles International Airport.[30] The Edward C. Little Water Recycling Facility produces five types of "designer" recycled water and is the largest water-recycling facility of its kind in the United States. The idea is to only treat water to the level needed for a given use, a concept commonly known as *fit for purpose*. Tertiary-treated water is used for irrigation. Wastewater specially processed to remove ammonia is used for industrial cooling towers. Two levels of treatment using reverse osmosis are used for low- and high-pressure boiler feed water for major refineries. Finally, a fifth type of water undergoes advanced water treatment and is injected into groundwater for a seawater barrier and to augment local well-water supplies.[31] Unlike the experiences in the Upper San Gabriel and East Valleys, the potable reuse has operated without controversy since 1995. It appears the third time's the charm, but there's another reason for the project's smooth sailing.

When people hear about the use of recycled water for a seawater barrier, they don't connect the dots with groundwater that they may ultimately consume. There's also a key difference in the purity of water required for spreading basins and water-well injection. Water that enters spreading basins undergoes *soil-aquifer treatment* as it slowly seeps through soils and shallow aquifers. Microbes digest contaminants along the way. Soil-aquifer treatment is considered a key part of the water-recycling process for water receiving only tertiary treatment and disinfection. In contrast, well injection of treated wastewater requires advanced water purification to protect the quality of groundwater, which has much less capacity to naturally cleanse itself. In addition, high-quality water is needed to prevent well clogging.

Despite the ups and downs, the city and county of Los Angeles have steadfastly pursued water recycling in a continuing drive to reduce or eliminate their dependence on imported water. Local sources of freshwater are cheaper and use less energy than imported water. Recycled water is also less vulnerable to climate change or interruptions by a catastrophic earthquake or severe drought. By 2020, the County Sanitation Districts had reused more than a trillion gallons of wastewater over the years. It's not just water that was saved. By pumping less water from northern California, the use of recycled water has avoided release of over 7.6 million tons of carbon dioxide and more than 7,700 tons of other air pollutants into the atmosphere since 1962.[32]

Despite population growth, the County Sanitation Districts have had increasingly less water to reuse in recent years because of water-conservation efforts, such as low-flow toilets and water-efficient washing machines. Effluent production at the County Sanitation Districts' facilities is now at levels last seen in the late 1960s.[33] The county also doesn't make use of all the water it recycles. For example, customers are not interested in recycled water for landscape irrigation when it's raining during February. To make year-round use of their recycled water, agencies in the region have turned their attention toward generation of additional advanced-treated water for groundwater recharge.

The Water Replenishment District of Southern California (WRD) was formed in 1959 for the purpose of protecting the groundwater resources of the LA Basin. The WRD is the largest groundwater agency in California, serving over four million residents in forty-three cities in southern Los Angeles County. Groundwater accounts for approximately half of the region's water supply; the other half comes from imported water.

The WRD protects the groundwater basins through recharge at the Montebello Forebay spreading basins, as well as water injection at seawater barrier wells along the coastline to keep the ocean from further contaminating the fresh groundwater aquifers. The WRD is also responsible for monitoring and testing the groundwater throughout the region.

Water for the Montebello Forebay spreading basins is a combination of captured stormwater, recycled water, and imported water. For decades, the WRD has been working to wean itself off imported water by encouraging conservation and working with the County Sanitation Districts to improve the quality and quantity of recycled water. In 2003, under the catchy slogan of WIN (Water Independence Now), the Water Replenishment District began ambitious plans to eliminate the need for *any* imported water for recharge at its spreading basins by capturing additional stormwater and increasing the amount of recycled water.[34]

In an interesting case of public-relations marketing, the original program name for WIN was the Water Independence Network. In 2008, a Special Projects in Design class from the University of Southern California undertook a student project to create artwork and logos around the agency's mission. When the class made their presentation to the WRD, they were emphatic that WIN should stand for Water Independence *Now* (replacing *Network* with *Now*) to convey that this was a timely and pressing goal and not just an inanimate network. The WRD readily accepted this sage advice.[35]

The cornerstone of WIN is the Albert Robles Center for Water Recycling & Environmental Learning, a state-of-the-art advanced wastewater-treatment plant adjacent to the Montebello Forebay spreading basins. This advanced-treated water is mixed with tertiary-treated wastewater from the County Sanitation Districts. In 2019, the Water Replenishment District achieved the WIN goal of using only local water (recycled water and stormwater) to recharge the LA Basin groundwater supply. Not a drop of imported water was used.

With this big WIN under its belt, the Water Replenishment District has been working on an even more ambitious program—Water Independence Now for All (WIN 4 All). Although the replenishment district has freed itself from using imported water for recharging spreading basins, the four million people in its service area still directly rely on imported water for half their water supply. The WRD has set its sights on developing a local sustainable water supply for the entire region with no imported water. The key is using the groundwater system as a subsurface reservoir. Lowered groundwater levels from the pumping over the years has freed up space for storing nearly a half million acre-feet of water underground. WIN 4 All depends on partnering with the county and city of Los Angeles to make use of the region's two largest untapped sources of wastewater: the Joint Water Pollution Control Plant and the Hyperion Water Reclamation Plant.

The Joint Water Pollution Control Plant (JWPCP) is the largest of Los Angeles County's eleven wastewater-treatment plants and one of the largest in the country. It's also the only wastewater-treatment plant of the eleven that does not produce recycled water. As the county's most downstream plant, the JWPCP discharges all its treated effluent to the Pacific Ocean. Adding advanced wastewater treatment at the JWPCP, together with conveyance structures to move the water "upstream" to spreading basins, has considerable potential but is also a monumental task. Enter the nation's largest wholesaler of treated water—the Metropolitan Water District of Southern California (MWD)—which is proposing to build, own, and operate the advanced treatment facility.[36]

MWD provides the imported water that the Water Replenishment District is trying to get away from using, and so under the general theme *if you can't beat them, join them*, MWD is getting into the recycling business. MWD also recognizes that underground storage provides resiliency during dry years when imported water is limited and the agency has trouble meeting all of its customers' demands for water.

Thinking big, MWD's goal is to produce 150 million gallons a day of recycled water, which would make it the largest water-recycling facility in the United States. Purified water would be delivered, via sixty miles of new pipelines, to spreading basins in Montebello Forebay, the San Gabriel Basin (a second chance for the Upper San Gabriel Valley, where the "toilet to tap" moniker originated), and Orange County for groundwater recharge and storage. In addition, some would go to seawater barriers and industrial facilities. When allowable at some future date, MWD's recycled water would also go to two of their existing drinking-water treatment plants, allowing the water to go directly into their massive water distribution system for direct potable reuse.

A demonstration project came online in 2019.[37] The project comes with a hefty price tag. Such a monumental project also would involve complex interagency agreements, extensive regulatory approvals, and considerable public outreach. In an interesting development, the Southern Nevada Water Authority might invest up to $750 million in the water-treatment project in return for some of MWD's share of Colorado River water.[38] Arizona is similarly interested.

The City of Los Angeles is also envisioned as a key partner in WIN 4 All. The city has not achieved the same level of water recycling as the county but has ambitious plans to catch up. In 2019, Mayor Eric Garcetti announced that Los Angeles plans to recycle 100 percent of its wastewater by 2035, as part of the L.A. Green New Deal that would establish the city as a national leader in renewable energy, environmental sustainability, and green jobs.[39]

There are many challenges to this lofty goal. The lion's share of the city's wastewater passes through the Hyperion Water Reclamation Plant, the largest wastewater treatment plant west of the Mississippi. For decades, the plant discharged primary-treated (at best) sewage into Santa Monica Bay, making surfers and swimmers sick and causing skin lesions on dolphins. It was not until 1998 that the Hyperion plant treated its wastewater to the full secondary levels required by the Clean Water Act before piping it to the ocean. Every day, the Hyperion plant discharges enough treated wastewater into the ocean to fill the Rose Bowl two and a half times over.[40]

Reuse of the city's wastewater has been a major focus of Los Angeles Waterkeeper. This environmental group strongly advocates for wastewater reuse in place of ocean desalination. Waterkeeper took on the State Water Resources Control Board (California's lead water-quality agency) under a relatively obscure provision of the California Constitution. Known as the Waste and Unreasonable Use Doctrine, it mandates that the state's water resources "be put to beneficial use to the fullest extent of which they are capable, and that the waste or unreasonable use or unreasonable method of use of water be prevented."[41]

In August 2020, the Los Angeles Superior Court ruled in favor of Waterkeeper, compelling the state board to analyze whether it is "wasteful" and "unreasonable" to dump billions of gallons of wastewater into the sea, when it could instead be used productively. Los Angeles Waterkeeper celebrated the ruling as a landmark win for more sustainable water management. "The days are numbered for the environmentally disastrous and economically costly practice of pumping water great distances over mountain ranges, using it once, and then basically throwing it away," declared Bruce Reznik, the group's executive director.[42]

As an interesting part of their decision, the court questioned whether money spent on water-conservation efforts would have been better spent on recycling wastewater discharge. The court pointed to a case where the city spent $500 million in rebates for homeowners to undertake desert planting in lieu of grass in their yards. "The benefits of this expenditure were dubious," the court wrote in the decision. "Could these monies have been better spent recycling the . . . wastewater discharge? We cannot know until the State Board conducts an evaluation of the reasonableness/waste of the discharges."[43]

Another challenge facing the city is the tradeoff between water for reuse versus water for the Los Angeles River. From its headwaters in the San Fernando Valley, the river travels fifty-one miles before emptying into the Pacific Ocean at the Long Beach waterfront. During dry weather, three of the city's wastewater treatment plants (not including Hyperion, which is close to the coast) are the main source of river water.

While historically often dry during the summer months, the river frequently became a raging torrent of water during the rainy season. A particularly devastating flood in 1938 killed ninety-six people and destroyed more than fifteen hundred homes. The floodwaters surrounded Warner Brothers Studios like a moat. So many Hollywood stars were stranded at their homes that the Academy Awards presentation was postponed for a week.[44]

The U.S. Army Corps of Engineers began straightening and encasing the river in a deep concrete channel to keep it from overflowing its banks during future floods. They erected fences and put up "No Trespassing" signs. A waterway that was once home to bears, deer, and steelhead trout, and shaded by lush alder, sycamore, and willow trees, was converted into perhaps the world's largest storm drain. While the concrete structure saved lives and prevented property damage, it also resulted in a river that was an eyesore and a public nuisance.

The turnaround is often credited to Lewis MacAdams, who founded Friends of the LA River in 1986. As the story goes, MacAdams and two friends, "with whiskey in their blood and wire cutters in their hands," cut a hole in the chain-link fence that bordered what had become an unsightly drainage ditch. They declared that the Los Angeles River was now "open for the people." The group foresaw a swimmable and fishable river accessible to everyone.[45]

Despite the drama, there wasn't an immediate groundswell of support for this idea. However, plans for river revitalization eventually began to take hold. Concrete was removed from large stretches of the river. Public parks and bike paths were built along its banks, encouraging recreational use. The river was declared a navigable water by the EPA in 2010, thus subject to the Clean Water Act. The following year, it opened to kayaking. The river now hosts a variety of riparian species that had lost most of their habitat to channelization and urban development.[46]

In one of those serendipitous connections, MacAdams was Garcetti's creative writing teacher in high school, influencing the mayor in making continued revitalization of the river a top priority of his administration.[47] With a dose of hyperbole, Garcetti calls the LA River "an iconic treasure, a place that holds a special place in the history of our city and limitless potential for the future of our communities."[48]

Given the challenges in balancing the two competing goals of water reuse and river revitalization, Mayor Garcetti's vision to recycle all of the city's wastewater by 2035 is likely an aspirational rather than realistic goal. Nonetheless, by collectively working on the problem, much can be achieved. As Lewis MacAdams liked to say, "If it's not impossible, I'm not interested."[49]

Around 2000, Earle Hartling and Bill Van Wagoner were invited to Orange County to talk about their experiences and lessons learned with water recycling. Hartling and Van Wagoner were battle-scarred. They had been called baby killers, confronted with kids holding signs that said "Please Don't Kill My Grandma," and faced with the antics of a verbally abusive clown. They had a simple message for the Orange County water and sanitation officials: *Don't worry about the money for outreach. Spend whatever you need in order to engage people and get everyone on board from the beginning. Work with the kids, because they get it (after they have some fun with potty talk). But* before proceeding any further, Hartling had one key piece of advice: "Reach around behind you and find your spine," he told them, "because you're going to need it."[50]

Chapter Four

Breakthrough

The Orange County Story

[We held] face-to-face talks for ten years to everybody that would listen.

—Ron Wildermuth[1]

As San Diego's potable-reuse plans were unraveling, an entirely different scenario was playing out in Orange County, its neighbor to the north. After more than a decade of planning and construction, the world's largest advanced wastewater-treatment plant for potable reuse came online in 2008 and was soon a global poster child for wastewater recycling.

Active opposition to the Orange County treatment plant was nowhere in sight. City, state, and federal officials were onboard. All major environmental groups backed the project, as did many health experts and medical doctors. The AARP, local chambers of commerce, and the Orange County Farm Bureau expressed their support. The Kiwanis, Rotary, and more than two hundred other community groups favored the project. A review of 158 newspaper articles about the project concluded that even the newspaper coverage was "notable in its utter lack of negative coverage."[2]

This was not the first successful potable-reuse project, but it represented a major breakthrough. In 1962, groundwater recharge of recycled water at Whittier Narrows had flown under the public's radar.

Such projects were viewed as the domain of "experts," and public engagement was not a major feature. Three decades later, the public was of a different mindset. Despite a multi-year drought in 1987–1992, the 1990s turned out to be a risky time to start up a potable-reuse project in southern California. As we've seen, three high-profile projects failed because of public outcry. In each case, officials were taken completely by surprise. In 1993, the water-reuse project in the San Gabriel Valley was challenged by Miller Brewing and abandoned a few years later. In 1999, San Diego's mayor and city council shut down the city's Water Repurification Project. In 2000, the San Fernando Valley project (East Valley) was shuttered just as it was ready to go online, after having spent $55 million. Given this checkered history, the highly acclaimed potable-reuse project in Orange County took on special meaning.

It's hard to overestimate the importance of groundwater to the 2.5 million residents who overlie the extension of the LA Basin into Orange County. Fortunately, the importance of this resource was recognized very early in the area's development. In 1933, the state legislature established the Orange County Water District (Water District) to manage the area's groundwater and protect the county's water rights to the Santa Ana River. Today, the Water District provides water for nineteen municipal and special water districts in north and central Orange County. Approximately three-quarters of this potable supply comes from the local groundwater basin.

When the Orange County Water District was formed, groundwater was pumped mostly for agriculture. Citrus trees and dairy farms dominated much of the landscape. Farmers also cultivated lima beans, celery, walnuts, and berries. The Knott's Berry Farm amusement park began as a roadside fruit stand on a working berry farm.[3] Disneyland would be carved out of Anaheim's orange groves. In the decades following World War II, the area rapidly transitioned from an agricultural center into a densely populated suburban area. Groundwater levels were dropping, as natural recharge could no longer offset withdrawals.

Soon after its formation, the Water District began acquiring portions of the Santa Ana River channel and adjacent lands in the upper part of the basin. Similar to the Whittier Narrows area in Los Angeles, these areas were well suited geologically for recharging the groundwater system. Purchasing these lands early on proved to be a smart move, in what is now a highly developed area. The first land was purchased for $28 an acre in 1936. A purchase in 2014 cost $1.6 million per acre.[4]

The Water District first turned to Santa Ana River water for groundwater replenishment, but the river flows were limited and fluctuated from year to year. As imported water became available from the Colorado River, the Water District began to supplement the local river water with imported water. Later it began to use water imported from northern California through the State Water Project. With almost two-thirds of California's population in southern California, where less than one-third of the state's precipitation falls, Governor Pat Brown proclaimed that the State Water Project would "correct an accident of people and geography."[5] It would prove only partly up to the task.

A major shot across the bow came in 1963, when the Supreme Court ruled in favor of Arizona on allocations of Colorado River water. More than half of southern California's former "entitlement" of Colorado River water was lost to Arizona. The original proposal for California's massive State Water Project also had been cut back, as plans to divert water from the Klamath and other north coast rivers were abandoned because of local opposition and concerns about effects on salmon. As a result, the State Water Project supplied only about half of the amount originally planned for by the Water District. These imported water supplies also were becoming more expensive, as well as vulnerable to interruption by droughts or a major earthquake.[6]

As the population continued to grow, increases in groundwater demands lowered the water table below sea level, causing seawater to move landward into the aquifers across a several-mile stretch of coast between Newport Beach and Huntington Beach. Known as the Talbert Gap, the seawater intrusion was the result of a gap that had been carved by the ancestral Santa Ana River and subsequently buried. Seawater intrusion was first detected in the 1930s. By the 1960s, seawater had intruded as far as five miles inland, forcing the closure of numerous municipal supply wells.[7]

Treated wastewater was seen as a logical candidate to help in the battle against seawater intrusion and groundwater depletion. As the population grew, wastewater was growing in volume and seen as virtually 100 percent dependable compared to the vagaries of imported water.

In the mid-1970s, Orange County completed a twenty-first-century, state-of-the-art treatment plant for recharging treated wastewater effluent. Water Factory 21 sounds like one of those dismal assembly-line factories out of the old Soviet era, but the truth was just the opposite. Water Factory 21 was the first major water-recycling plant in the world to use reverse osmosis to purify wastewater to drinking water standards. The treated water was blended with deep well water and injected into a series of injection wells to create a hydraulic barrier at Talbert Gap. The majority of this water flowed into the groundwater basin to augment Orange County's groundwater supply—a two-for-one benefit.[8]

Reverse osmosis takes its name from *osmosis*, a process first observed by French physicist Jean-Antoine Nollet in 1748. Osmosis is an important process in biological systems, causing water to migrate through a membrane from an area of low salt concentration (as in a plant's stem) to one of high concentration (like the interior of the plant's cells). This process works to equalize the concentrations on each side of the membrane. As the name suggests, reverse osmosis (RO) turns this concept on its head by using high-pressure pumps to force water molecules across a membrane that has tiny gaps just slightly larger than a water molecule. Anything larger, as in most dissolved substances and pathogens, is blocked by the membrane. Purified water (permeate) is sent in one direction and the reject stream (concentrate) in another.

In 1959, Sidney Loeb and Srinivasa Sourirajan, graduate students at the University of California at Los Angeles, developed the first commercially viable RO membrane. Research on desalination had begun under the Eisenhower administration and increased under President John F. Kennedy. Five pilot plants were built to test different desalination methodologies.[9] When the first of these opened in 1961 in Freeport, Texas, Kennedy enthusiastically proclaimed: "This is a work which in many ways is more important than any other scientific enterprise in which this country is now engaged."[10]

One of the five pilot plants was built for the Navy on the Point Loma peninsula in San Diego, but its life there was short-lived. In 1964, Fidel Castro threatened to cut off the water supply to Guantanamo Bay. In response, the San Diego plant was taken apart and reassembled at the Guantanamo Bay naval base.[11]

Reverse osmosis is one of those cases where a technology developed for one purpose (desalination) was discovered to be very useful for another (removing all sorts of contaminants from water). Since its first application at Water Factory 21, RO technology has been used throughout the world to clean wastewater and is currently required for all potable-reuse projects that use subsurface injection wells in California. Water Factory 21 also set the stage for Orange County's next, and even more impressive, venture with potable reuse.

By the 1990s, it was clear that the Water District needed more water to address continuing threats of seawater intrusion and groundwater depletion. At the same time, the Orange County Sanitation District, which provided the secondary-treated wastewater for Water Factory 21, faced the possibility of having to build a second ocean outfall that would cost approximately $200 million. And so, the two agencies put their heads together and came up with the perfect solution: purify the wastewater instead of sending it to the outfall. As an added plus, this approach would provide cheaper drinking water than either importing water or desalting seawater. The concept of a larger and more advanced water-reuse facility began to take shape.

In January 2008, after more than a decade and $481 million to complete, the world's largest advanced water-treatment system for potable reuse opened for business. Known by the unassuming name *Groundwater Replenishment System*, the facility produced five times more purified wastewater than Water Factory 21.[12] The timing couldn't have been better, as California was in the midst of a serious multi-year drought. "It made us look like geniuses," Water District general manager Mike Markus later told the *New York Times*.[13]

The Groundwater Replenishment System (GWRS) purifies secondary-treated wastewater through a three-step advanced-treatment process. The first step, microfiltration, uses polypropylene hollow fibers about the thickness of dental floss and with tiny holes in the sides 1/300th the diameter of a human hair. By drawing water through the holes into the center of the fibers, suspended solids, protozoa, bacteria, and some viruses are filtered out of the water. To prevent clogging, each microfiltration cell is backwashed every twenty-two minutes and undergoes a full chemical cleaning every twenty-one days.[14] The Orange County Water District was a pioneer in using microfiltration for potable reuse.

Reverse osmosis is the second step, where pretty much everything but the water molecules are removed. (There are a few notable exceptions, as we'll see.) Finally, the water is dosed with hydrogen peroxide and then zapped with ultraviolet light—known as the advanced oxidation processes. This third step further disinfects the water and destroys trace organic chemicals that may have snuck through the RO membranes.

While the advanced water treatment gets all the attention, post-treatment of the water is also extremely important. The end result is water so pure that minerals must be added back to be healthy to drink. Also, without adjustment, the water is highly corrosive to the concrete pipes. Additional fine-tuning is sometimes required to minimize mobilization of arsenic that naturally occurs in certain aquifer rocks.[15]

As of June 2021, the Groundwater Replenishment System has treated 350 billion gallons of water for reuse.[16] About 30 percent of the water is pumped into injection wells, where it serves as a barrier to seawater intrusion. The rest is piped thirteen miles to spreading basins not far from Disneyland, where the water filters through the sand and gravel, replenishing the groundwater system. The process uses less than half the energy it takes to transport water from northern California, and less than a third of the energy required for desalination of seawater.[17] Although the Water District does not have to use RO for all of the water applied to the spreading basins (i.e., undergoing soil-aquifer treatment), it does so to assure the public that it is taking every precaution to keep the groundwater safe.

Expanding the GWRS to one hundred million gallons per day in 2015 made the world's largest advanced water-purification system for potable reuse even bigger. A final expansion is scheduled for completion by 2023, bringing the total production to 130 million gallons per day—enough water for a million people. At this point, Orange County will run out of additional wastewater to recycle.[18]

The GWRS has garnered more than fifty awards and is widely regarded throughout the world. It was declared "Officially Amazing" when it set the Guinness World Records title for the most wastewater recycled to drinking water in twenty-four hours on February 16, 2018.[19]

The success of the GWRS did not just happen. The Orange County Water District undertook an impressive and highly proactive public-relations campaign that began in earnest ten years before the project came online. Along the way, unexpected problems were dealt with in a very transparent manner.

Ron Wildermuth was hired to run Orange County's early outreach effort. Wildermuth was a retired Navy captain who later served as the public affairs officer for General Norman Schwarzkopf during the Persian Gulf War. He came onboard about the time San Diego's water-purification plan fell apart. "It was such a disaster," said Philip Anthony, the Water District board president at the time. "We were really concerned about the public relations."[20]

The communications strategy was extremely diverse. They focused on face-to-face meetings with influential residents, while also taking their plan to the community via neighborhood pizza parties, water-treatment plant tours, and hundreds of public meetings where they explained how sewer water would be purified and then added to underground water supplies. "We talked to the historical society. We talked to the chambers. We talked to the flower committee. If there was a group, we talked to them," Wildermuth recalled.[21] Another smart move was community polling on a regular basis, which allowed them to track the impact of outreach efforts.

The Orange County Water District not only didn't try to dodge the "yuck" factor, but they also put it right up front. "We started telling people from the start that we're purifying sewage water," said Wildermuth. "We have not had a group oppose the project after they've listened to the project and the alternatives."[22] Public television personality Huell Howser was hired to narrate a video explaining how earthy-smelling wastewater will be transformed into distilled, crystal-clear water.

Minority outreach was prioritized with the county's very large Hispanic and Vietnamese communities.[23] Many people in these communities have a basic mistrust of public water systems because their home countries had such a poor record of providing safe water. Tours and a technical brochure were available in Spanish, Korean, Chinese, and Vietnamese. A steady stream of visitors from around the globe toured the project.

The kids were also key. The country's largest water education festival was developed in Orange County for third-, fourth-, and fifth-grade students to educate them about local water issues and how they can help protect water supplies and the environment.

Accommodating future population growth was part of the Water District's early messaging but was soon deemphasized when they realized people were concerned about urban sprawl. Therefore, they shifted their focus to the cost of the recycled water compared to other options, as well as the reliability of potable reuse during times of drought. Drought resiliency became a prominent talking point, but Orange County staff stressed that potable reuse shouldn't be viewed as a last resort, but rather as part of a diversified water portfolio.[24]

In 2017, the Water District became the first in the Western Hemisphere to bottle purified wastewater for educational purposes. The bottled water allows people throughout California to see it, taste it, and get comfortable with it.

In 2000, a probable human carcinogen known as N-nitroso dimethylamine (NDMA for short) was found in the water being injected into the seawater barrier at Talbert Gap and in water pumped from nearby public-supply wells. This discovery had the potential to derail the proposed GWRS project—not to mention the Orange County Water District's reputation.

NDMA and other nitrosamines first received attention in the 1970s in connection with processed foods (notably cured meats such as bacon) and beverages such as beer and milk. NDMA wasn't found in drinking water or wastewater until the late 1990s, when analytical methods improved to where it could be detected at the parts-per-trillion levels. The U.S. Environmental Protection Agency estimates that drinking water with approximately 0.7 parts per trillion NDMA results in a one-in-one-million cancer risk.[25]

From the beginning of Orange County Water District's recycling project, officials had reassured the public that transparency would be the cornerstone of all communications. To maintain trust, it was critical to be the first to communicate any problems, be factual, and not hold back bad news. Finding NDMA in their water put that assertion to the test. The assistant general manager later described the Water District response:

> Some of us on the water quality end of the business wanted to get answers to the problem. See what we can do to fix it, first. [The public relations specialist] said no, that we needed to talk to the public, we needed to actually call the media in and do press briefings. . . . His instincts were right. If the media and the public perceive you as having nothing to hide, if you've got something that goes wrong, you're going to tell them about it.[26]

The Water District staff quickly began preparing a communications plan, issued a press release, set up a toll-free hotline, and invited members of the local media to meet and discuss the findings. This proactive approach included the disclosure of the test results and the actions the Water District and Sanitation District were taking to reduce NDMA in water from Water Factory 21 and the proposed GWRS project. Part of the problem was a metal finishing plant that was discharging NDMA precursors.[27] The plant immediately changed processes.

One production well had concentrations of NDMA above the "notification level" at which the state recommends notifying customers about the chemical and associated health concerns. No other action is required. Instead of simply being the bearer of bad news, the Water District installed and operated a UV treatment system on the well for nine years until the NDMA levels were consistently below the analytical detection limit.[28]

The rapid outreach effort resulted in balanced stories in the news and no public or political backlash. By responding decisively, the agencies also had more credibility in explaining that people are likely more exposed to NDMA from hot dogs and beer than from their drinking water. The finding of NDMA was instrumental in the decision to add UV light to the GWRS treatment train. UV is effective at destroying NDMA, whereas RO is not.

NDMA wasn't the only challenge. In 2002, the Orange County Water District detected 1,4-dioxane, a suspected human carcinogen, in groundwater near the Talbert Barrier at levels that exceeded the state notification level of three parts per billion. Once again, the Water District took immediate action. Nine production wells were temporarily shut down, with a loss of thirty-four million gallons per day of water supply. An investigation traced the contaminant to a single entity that discharged 1,4-dioxane into the sewer system. The discharger voluntarily ceased discharging 1,4-dioxane to the sewer, the concentrations declined, and the wells were returned to service. As a future prevention, the advanced oxidation process, which is effective in treating 1,4-dioxane and similar contaminants, became part of the GWRS treatment train.[29]

Today, the Orange County Water District tests for more than five hundred compounds, many more than required by state and federal regulations. It tests water from approximately 1,500 locations throughout the basin, analyzes more than 20,000 samples each year, and reports more than 400,000 results. The Water District also provides regional testing of more than two hundred drinking water wells for local providers.[30]

The Orange County experience demonstrates a commitment to innovation in treatment technology, continuing efforts to go well beyond minimum permitting requirements, and a proactive and transparent response to any problems that arise. As Mike Markus, the Water District's general manager, puts it: "As my public information officer told me early in the project, 'Mike, this is not an engineering project, it's a PR project.'"[31]

Chapter Five

From Toilet-to-Tap to Pure Water

A crisis is a terrible thing to waste.

—Paul Romer, Stanford University economist[1]

In 2004, five years after the San Diego City Council forbid the use of city funds or staff toward water purification, Mayor Dick Murphy and the city council voted unanimously to allocate one million dollars for a Water Reuse Study.[2] Both potable and nonpotable uses would be considered. The change of heart once again began with controversies over the Point Loma wastewater-treatment plant.

When the U.S. Environmental Protection Agency (EPA) renewed the Point Loma plant waiver in 2001, environmental groups mounted lawsuits challenging the decision. After three years of negotiations, the environmentalists dropped their opposition to the waiver. In return, the city agreed to the Water Reuse Study, as well as to evaluate an improved ocean-monitoring program and test new treatment technology at the Point Loma plant.[3]

Over the next decade, the Point Loma plant would continue to loom in the background. The Carlsbad desalination plant would be built in north San Diego County and be viewed as competitive with potable reuse.[4] And a subsequent mayor and some city council members would fight the reuse project. This time, however, events would turn out differently. A series of severe droughts highlighted the need for water reuse. Success also can be credited to a host of community leaders.

San Diego's renewed effort built off the recommendations of a 2003 report by a state task force on water recycling. The report stressed that any decision to undertake potable reuse "needs to be a local decision based on community values."[5] It emphasized the following: Involve the public in all phases of project planning; don't just inform them of final decisions. Listen and respond to public fears and concerns, and if necessary, mitigate them with changes in project design. Disseminate understandable information in many forums. Incorporate principles of environmental justice. Provide the public with a broad understanding of water-supply issues so that they have a context in which to evaluate the need for recycled water.

The San Diego potable-reuse project had previously followed much of this guidance, but in one major change, this time the public would be engaged from the get-go in developing and selecting alternatives. The American Assembly process developed by President Dwight D. Eisenhower more than fifty years earlier served as the model.

San Diego's sixty-seven assembly participants were selected through a city-wide search for interested community leaders and professionals in various sectors. The mayor and city council members suggested names of constituents to participate. Potential candidates were contacted, provided with an overview of the study and process, and asked if they would commit to an active role.[6]

The assembly, together with a group of experts in water reuse, developed and evaluated six options for increased water recycling. The options included those with and without potable reuse. At the end of the process, the assembly unanimously backed the same option considered in the 1990s—indirect potable reuse using reservoir water augmentation. Recycled water would undergo advanced treatment and be sent to a reservoir, where it would blend with other sources of water, and then that mix would be treated again before being delivered to customers citywide. The group's selection of this as the preferred alternative was a major coup for moving forward.

A key advantage of the reservoir water augmentation option is that it would reuse the greatest amount of wastewater. The city had built two wastewater-reclamation plants for nonpotable reuse, one north of the city and one south. The two plants met the Ocean Pollution Reduction Act requirement for a combined recycled-water capacity of forty-five million gallons per day.[7] However, capacity is one thing—actual use is another.

The market for nonpotable reuse water had failed to take off. Basically, it's not easy to retrofit a developed area for nonpotable use. Building the purple pipe network and connections to users was expensive, the water was too saline for some applications, and water demand to irrigate landscaping (the main use) was seasonal, dropping precipitously during the winter. The two reclamation plants produced recycled water at about 15 percent of overall capacity.[8] The remainder was treated to secondary standards and discharged back into the sewer system, where it mixed with raw sewage, only to be treated again downstream at the Point Loma plant before being dumped into the ocean.[9]

Despite the need for potable reuse to meet the city's water-reuse commitment, the proposed potable-reuse project soon faced two major opponents: a new mayor and the daily newspaper, the *San Diego Union-Tribune*.

Mayor Dick Murphy's fate has been compared to the captain of the *Titanic*—a man remembered for a journey cut short by something hidden beneath the surface that he never saw coming.[10] In Murphy's case, the surprise was a funding crisis in the city's pension system. The problems began before Murphy took office but ballooned during his tenure. As the city struggled through the financial crisis, San Diego became known as "Enron by the Sea."

Murphy was up for reelection in 2004. Councilwoman Donna Frye, a vigorous advocate of environmental protection and strong proponent of the potable-reuse project, launched a last-minute write-in campaign in the mayoral race. Frye garnered the most votes, only to have a judge rule that several thousand were invalid because they misspelled her name. Murphy was declared the victor. But it was a short-term victory. Murphy resigned the following April as the financial crisis widened.

Jerry Sanders, San Diego's former police chief, faced off against Frye in a special election for mayor. After coming in a distant second in an eleven-person primary race, Sanders beat Frye in a November runoff. Sanders viewed potable reuse as a political trap and was antagonistic toward environmentalists. He strongly favored desalination, which he viewed as competing with water recycling.

In 2006, the Water Reuse Study results came before the city council. The study had been completed before the mayoral race but was held back so as not to repeat the 1990s fiasco of getting caught up in election politics. Sanders quickly announced his opposition to any further work on potable reuse, saying that although he didn't dispute the science, the project was "expensive, divisive, and unnecessary."[11] A few days later, the *San Diego Union-Tribune* joined in against the project. "Your golden retriever may drink out of the toilet with no ill effects. But that doesn't mean humans should do the same," the newspaper's editorial began.[12] The paper repeated the falsehoods that some of San Diego's poorest neighborhoods were targeted for the treated wastewater and that desalination would be a much less expensive alternative.

An advocacy group known as the Water Reliability Coalition pushed back, promoting potable reuse. This was a coalition of odd bedfellows. Among its several dozen members were the Audubon Society, Building Industry Association, Chamber of Commerce, Friends of Infrastructure, and Surfrider Foundation. "Early on, someone suggested that we call ourselves the Unprecedented Coalition because of the diverse membership," quipped Lani Lutar, head of the San Diego County Taxpayers Association. Lutar, a former intern at Coastkeeper, played a lead role in bringing pro-business groups into the coalition.[13]

In October 2007, the San Diego City Council voted to initiate a pilot project to evaluate potable reuse. Despite lobbying by the Water Reliability Coalition, Sanders vetoed the project. The city council promptly overrode Sanders's veto, but the mayor refused to authorize funds. After a year of this back-and-forth, the city temporarily raised sewer fees to provide the $11 million needed for the pilot project.

Meanwhile, the Point Loma plant's waiver was becoming more and more difficult to renew. San Diego was the last large waiver holder in the United States. The city had been warned by the State Water Resources Control Board that it "should not expect to receive waivers forever."[14] At the same time, a study conducted by Scripps Institution of Oceanography (as part of the 2004 settlement agreement between the city and environmental groups) found no evidence that the plant effluent endangered the marine environment.[15]

During the 2008–2010 waiver application process, local environmentalists would again play a pivotal role. Marsi Steirer, former deputy director of the Public Utilities Department, recalls that the environmentalists were a force to be reckoned with who proved to be trustworthy partners in the long run. Steirer considers herself "the mother of San Diego's potable reuse demonstration project," which people involved with the project say is by no means an overstatement. Steirer and her team managed to hang in there throughout years of ups and downs and political infighting—oftentimes a daunting task.[16]

A turning point came when Marco Gonzalez, attorney for Coast-keeper and the local chapter of Surfrider Foundation, began to rethink the environmentalist's single-minded focus on opposing San Diego's waiver. Gonzalez says his epiphany came while sitting on a surfboard one summer evening in 2008. He recognized that their goal was to mini-mize (or eliminate) wastewater discharges to the ocean *and* maximize water reuse—and accomplish both as soon as possible. If the city agreed to the treatment plant updates, it would take a decade or longer to fully implement them. Plus, the expensive upgrade would compete with funds needed for new pipes and pumps to reduce sewer spills causing beach closures—a key priority for environmental groups and the focus of a second lawsuit against the city. It had become obvious that the city's already-stressed finances couldn't afford everything environmentalists wanted. Moreover, achieving full secondary treatment at Point Loma would turn into the biggest argument against constructing new water-recycling plants for decades to come until the loans were paid off. "We understood that despite our best efforts to secure federal and state assis-tance, money wasn't going to be growing on trees," Gonzalez says.[17]

And then there was the other major problem that wasn't going to go away. Upgrading the Point Loma plant would have no impact on San Diego's demand for water imported from hundreds of miles away or fight against the proposed desalination plant, which environmentalists vehemently opposed. Gonzalez realized that, rather than sink a fortune into secondary treatment at Point Loma and making the discharge slightly cleaner, the city should be reducing its total ocean discharge through recycling. He was soon joined by Bruce Reznik, executive director of Coastkeeper.

Instead of shying away from the "Toilet-to-Tap" moniker, Reznik went on an awareness campaign to educate the public and city officials that recycled water was already a normal part of their lives. He came up with a clever one-liner, "What happens in Vegas doesn't stay in Vegas," by way of explaining that the Colorado River water that San Diegans drink comes to the city with plenty of partially treated sewage from Las Vegas and more than two hundred other municipalities.[18]

In February 2009, San Diego reached a landmark cooperative agree-ment with Coastkeeper and Surfrider when the two environmental groups agreed not to oppose the city's latest waiver for the Point Loma plant. For its part of the deal, the city would conduct a two-year Recycled Water Study, looking beyond the forty-five million gallons per day to meet the waiver. This time, water recycling would be limited only by the amount of wastewater available to reuse. The result would be a new vision for water reuse in the San Diego region.

Coastkeeper and Surfrider agreed to use their best efforts to gain commitments from other environmental organizations and individuals not to oppose the permit. Many local environmental groups got on board. Others accused them of selling out. Gonzalez and Reznik weathered the storm and also helped the city diffuse concerns by the California Coastal Commission, which had opposed the previous waiver.

Mayor Sanders continued to oppose potable reuse but was under increasing pressure to switch sides. In 2009, with California entering the third year of drought, biotechnology executives met with the mayor to underscore that water shortages posed a threat to their businesses. Biotechnology is a key industry in San Diego, and it requires a dependable supply of high-quality water. "They were talking about moving away from San Diego," Sanders later told a *New York Times* reporter.[19]

In early 2010, Sanders half-heartedly told a local news organization that he supported the pilot project. "My concern has been and will always be that that water is safe. We're still concerned about the pharmaceutical uses, but I'm certainly not going to quibble with scientists and demagogue this issue."[20] In June 2011, when Sanders kicked off a one-year test of the pilot facility, he told onlookers, "San Diego has elected to move beyond its fear and let science do its talking," When asked if he would take a drink from a beaker of treated wastewater in front of him. "Nooooo," the mayor replied, with a nervous laugh. "Why not?" a reporter asked. "Actually, they won't let you," Sanders replied, but he was clearly very uncomfortable about the prospect.[21]

Meanwhile, the *San Diego Union-Tribune*'s position had begun to shift after the opinion editor, Bob Kittle, an ally of Sanders, left the paper in 2009.[22] In 2011, the newspaper reversed its position with an editorial titled, "The Yuck Factor: Get Over It."[23]

The potable-reuse project was now moving forward on multiple fronts. A demonstration project opened to the public. An independent advisory panel was convened to provide expert peer review and feedback. A study was launched to evaluate water residence times and mixing in the reservoir targeted for augmentation. A regulatory framework was proposed. Energy and economic analyses were undertaken. And extensive education and outreach programs were developed.

In 2012, the Recycled Water Study (from the 2009 settlement agreement with environmentalists) concluded that San Diego could ramp up indirect potable reuse to eighty-three million gallons per day, which would provide about a third of the city's water supply. Costs of the recycled water were comparable to the current delivery costs for imported water and projected to be more economical in the long run as the price of imported water increased.[24] As an added bonus, the study noted that appliances and water heaters would last longer because of the reduced salinity from reverse osmosis.[25]

Public support for potable reuse increased from 26 percent in 2004 to 73 percent in 2013.[26] In 2014, with California enduring the hottest and most severe drought in the state's history, the San Diego City Council unanimously adopted a resolution supporting the potable-reuse project. Mayor Faulconer proclaimed, "Our city is presented with an incredible opportunity—to gain water independence, the ability to control our own water supply for the very first time." Councilwoman Marti Emerald added, "We can no longer afford to use water just once in this region. If we don't act today, it's literally kicking the problem down the road."[27]

Emerald was the television consumer-affairs reporter who brought the term *toilet-to-tap* to San Diego's attention in the late 1990s during her "sweeps week" television series. Now as a member of city council, she was fully onboard with the Pure Water program.[28] Faulconer had opposed potable reuse during much of his time on the city council. His turnaround came after he accepted an offer by Bruce Reznik to arrange a tour of Orange County's recycling plant for skeptical city council members. As a traditional Republican, Faulconer asked local business owners who were taking the tour why they supported potable reuse. "It just makes dollars and sense," came the reply. That was a good enough reason for Faulconer, and he soon became a strong advocate.[29]

The city still had a major concern. What would keep environmental groups from changing their mind when they were under new leadership in the future? After having spent billions on water purification, the city could find itself once again on the hook for a costly upgrade to its Point Loma plant.

In December 2014, a third cooperative agreement was reached with the environmental groups.[30] They agreed to get off the city's case on plant upgrades in return for a commitment to continue progress toward recycling wastewater. As part of the deal, the environmental groups supported federal legislation that would eliminate the need for Point Loma plant upgrades. Known as the Ocean Pollution Reduction Act II (OPRA II), it would grant the Point Loma plant *secondary equivalency*—essentially pollution credits for reducing its wastewater discharges as a result of the Pure Water project. The act overwhelmingly passed the U.S. House of Representatives in 2020 but has not been taken up by the Senate.

Since June 2011, San Diego has produced a million gallons of purified water daily at its Water Purification Demonstration Facility. Serving as both a testing and educational center, the Pure Water team calls the facility "the gift that keeps on giving." By 2020, almost nineteen thousand visitors had taken the free tour. Seeing and tasting is believing. Those who taste the water almost invariably come away satisfied. Unfortunately, tours became virtual during the COVID-19 pandemic.[31]

After more than twenty-five years of off-and-on preparation, San Diego is finally constructing the first phase of Pure Water San Diego to produce thirty million gallons a day of potable water by 2025. The advanced water-treatment facility will be located adjacent to the north city water-reclamation plant originally built for nonpotable reuse.

While Orange County has been on the cutting edge and is often held up as the poster child of potable reuse, it has several intrinsic advantages over San Diego. One is politics. The Orange County Water District was established as a special water district to manage the regional groundwater basin. A mayor or city council does not have to approve the program. Orange County also had fewer infrastructure challenges, as well as a large groundwater basin for storing the water.

Among the infrastructure challenges for Pure Water San Diego, the first phase requires an eleven-mile pipeline to transport additional wastewater to the north city plant and a second parallel pipeline to transport brine from reverse osmosis back to the Point Loma plant. Much of the current controversy about the Pure Water project (and a focus of city outreach) involves concerns by affected residents about the disruption of their neighborhoods during pipeline construction.

Lacking a suitable groundwater basin to serve as an environmental buffer, San Diego is turning to surface-water augmentation—the first such potable-reuse project in California. The city's largest reservoir, San Vicente, was the original choice but was later replaced by the much closer Lake Miramar. The switch saves money and reduces disruptions during pipeline construction, but also presents a major challenge. The storage capacity of Lake Miramar is about one-fortieth that of San Vicente Reservoir, which means less dilution and retention time.

As part of plans to compensate for the smaller reservoir, Pure Water San Diego has added ozone and biologically activated carbon filtration in front of the standard advanced-treatment train of microfiltration, reverse osmosis, and UV/advanced oxidation. The purified water will be released into Lake Miramar through an underwater pipeline, where it will blend with imported and local water supplies. Water withdrawn from the lake will be treated again before delivery to customers. Coincidentally, the Miramar plant serves the north part of San Diego, which will avoid a repeat of controversies about sending treated sewer water to poorer neighborhoods in the south.

Phase II planning is underway for additional facilities to bring the total amount of potable reuse to eighty-three million gallons a day by 2035. When fully operational, potable reuse could comprise nearly half of San Diego's water use. The increased percentage over the previous estimate of one-third is not due to more planned recycled water but rather because of reduced projections of water use as a result of conservation.

The enhanced water security that Pure Water San Diego brings comes at a price. Like many, if not most, such projects, cost estimates keep going up—from $3 billion when the city council first approved the concept in 2014 to $5 billion today.[32] But you can't put a price tag on water security, particularly in the southwestern United States. As a result of the city's tenacity, it's quite possible that San Diego will one day be held up as a poster child for water reuse.

Interest in potable reuse in San Diego County extends beyond the city of San Diego. In North County, the city of Oceanside is planning to meet about 30 percent of its water supply through potable reuse. Most of Oceanside's water is imported, but a small portion comes from the local Mission Basin Aquifer, which has declined over years of use. The city is planning to use the aquifer for potable reuse, simultaneously increasing the city's local water supply and improving the groundwater resources.[33]

East of San Diego, the Padre Dam Municipal Water District (the former Santee water district under Ray Stoyer) is pursuing its own advanced water-purification program. The goal is to provide up to 30 percent of drinking water demands for East County residents.[34] It also would eliminate the need to send most of East County's wastewater to the Point Loma wastewater-treatment plant.[35] The project is proceeding smoothly today, but the original proposal got off to a rocky start.

In 2005, the Helix Water District, which adjoins the Padre Dam water district, proposed to treat some of Padre Dam's wastewater with advanced treatment technologies and recharge groundwater in the nearby El Monte Valley for indirect potable reuse.[36] To offset the costs of the project, a company (curiously named El Monte Nature Preserve) would mine sand and gravel for aggregate—a high-demand resource that's locally in short supply. The mining would take place over a decade or so, followed by revegetation for riparian habitat and recreational trails (hence the name of the company: "Nature Preserve").

In contrast to Ray Stoyer's efforts with Santee Lakes, the public meetings were contentious with vocal opposition by local residents. The objections revolved less around the use of treated wastewater for drinking water and more around the mining impacts, including noise and traffic from mining operations, air pollution, and loss of domestic well water. Not everyone opposed it. The San Diego chapter of the Surfrider Foundation—whose members didn't live in the area—praised the project.[37]

In 2011, the Helix Water District's Board of Directors killed the potable-reuse project, citing increased project costs and delays in the availability of recycled water.[38] Plans for potable reuse were resurrected a few years later, this time using a nearby lake for surface-water augmentation. The partnership grew to include the Padre Dam Municipal Water District, Helix Water District, county of San Diego, and city of El Cajon. The re-envisioned project has been very successful at obtaining grants and loans and is scheduled to provide purified water to East County residents in 2025, the same year that San Diego's Phase I potable-reuse project using Lake Miramar is planned to open.[39]

Chapter Six

Colorado's Front Range

People seeing the beauty of this valley will want to stay, and their staying will be the undoing of the beauty.

—Chief Niwot upon meeting the first gold seekers
to visit Boulder Valley[1]

In the 1980s, Denver and surrounding communities were banking on a massive dam to meet their future water needs. The idea had been a dream of Denver-area officials off and on for decades. The Two Forks Dam would be built on the South Platte River about twenty-five miles southwest of Denver, near the river's confluence with its north fork. Almost as big as Hoover Dam, it would flood six towns and turn a popular wilderness area into the largest lake in Colorado.

Among the reaches inundated would be Cheesman Canyon, considered by many to be one of the world's best trout fisheries. The superselective rainbow and brown trout residing in this boulder-strewn river reach are legendary for their challenges. Experienced anglers claim that if you have the technique to catch fish in Cheesman Canyon, you can fool fish anywhere in the world. Despite the environmental impacts, Denver and forty metro-area suburbs were intent on building the dam. They were even willing to finance the billion-dollar expense themselves without any federal dollars.

Developers and municipalities led by the Denver Water Department (Denver Water) were pitted against environmentalists, outdoors enthusiasts, and those on the Western Slope who adamantly opposed plans by the Front Range cities to take more water from the Colorado River headwaters and transport it through a tunnel under the continental divide. The crowds were boisterous at hearings held in the two Western Slope counties where Denver would collect the water to fill the reservoir. A banner reading "Damn the Denver Water Board Instead" greeted attendees at one meeting. When asked for a show of hands of people opposed to Two Forks, the entire crowd jumped to its feet with whistles and catcalls.[2]

Despite the protests, Two Forks seemed a done deal when the U.S. Army Corps of Engineers announced in 1989 that it planned to issue a permit for the dam. But there was another key agency involved. The Clean Water Act grants the U.S. Environmental Protection Agency (EPA) veto power over such projects if the environmental impacts are unacceptable. The EPA regional office in Denver was split. The regional administrator favored the dam, whereas the EPA staff who reviewed the project opposed it. After EPA administrator William Reilly turned oversight of the review over to the regional administrator in Atlanta, the EPA regional office recommended against approving the Two Forks Dam. Reilly followed by killing the project in November 1990, a decision that essentially ended the era of big dam construction projects in the United States.[3]

In an amusing anecdote, Reilly and a friend visited Cheesman Canyon about a year after his veto, where they helped a fisherman push a truck out of a snowbank. Noticing that the truck had a "No Two Forks" bumper sticker, Reilly's friend told the fisherman that Reilly was the person responsible for killing the dam. "No (expletive)," replied the fisherman. "I've got some really great (expletive, nickname for marijuana) up at my house that I only break out for special occasions. Wanna go there and party?" The EPA administrator declined the invitation but recalled later that "it made me feel pretty good about my decision."[4]

The Two Forks decision had a major impact on Denver Water and other Front Range water utilities. "Collaboration and environmental stewardship would now move to the forefront of every project we do," recalls a Denver Water official.[5] Chips Barry, the utility's incoming manager, was blunter, later saying that Denver Water had to "quit fighting with everybody" and "make some friends."[6] In that spirit, the utility began to take a harder look at conservation and water reuse. There was plenty of opportunity for the former, as Denver had one of the highest per-capita water uses among major U.S. cities.[7] Water meters became mandatory for all Denver Water customers, and tiered water rates were instituted, with the unit price of water increasing with usage.[8]

During the decade following the Two Forks veto by Reilly, Colorado experienced relatively wet years, masking its underlying water-supply problems. The situation changed dramatically in 2002, with the onset of the worst drought in Colorado's recorded history. By September, Denver Water's reservoirs were about half full.[9] The drought also brought another threat to Denver's water supply.

On June 8, 2002, a Forest Service worker reported that a campfire in the mountains southwest of Denver was burning out of control. Before it was over, the Hayman Fire burned 138,000 acres, destroyed 133 homes, sent more than five thousand people fleeing for their lives, and led to six fatalities. It took nearly a month just to contain it. The fire was too dangerous to put fire crews on the front lines, and fire retardants dropped from planes evaporated before they hit the ground. At the time, the Hayman Fire was the most devastating wildfire in Colorado history.[10]

Authorities offered a reward for anyone with information about the camper(s) who had started the fire. But that was a false trail. There was no evidence of camping near the makeshift campfire ring where the fire began. Forest Service investigators also wondered why anyone would start a fire on a 90-degree day during a total ban on burning. As the investigators tried to reconstruct the incident, the Forest Service worker's story fell apart. She confessed to starting the fire, claiming that it was an accident from burning a letter from her estranged husband. Others believe the fire was meant to be a setup so she could become a hero by extinguishing it.

Whatever the reason, Denver Water continues to deal with the consequences. The fire burned most of the eighty-five hundred forested acres surrounding Denver Water's Cheesman Reservoir. An aerial photo of the reservoir taken after the fire shows a charred landscape and dead trees stretching as far as the eye can see.[11] Not to be confused with Cheesman Canyon further downstream, Cheesman Reservoir was Denver's first reservoir. When completed in 1905, the 221-foot dam was the world's highest and hailed as the solution to Denver's water-storage problems.[12]

When the fire was over, the area faced the longer-term threats of flooding and soil erosion. Vegetation is the key factor that limits the rates of erosion in most areas. Not only was the vegetation wiped out, but also the extreme heat from the fire impacted the ability of the soil to absorb and retain moisture.

In the aftermath of this and an earlier fire, Denver Water spent more than $27 million on sediment removal and water-quality improvements. Volunteer groups planted tens of thousands of trees, and Denver Water partnered with the U.S. Forest Service program From Forests to Faucets to maintain healthy forests in watersheds at greatest risk by thinning, removing underbrush, and educating homeowners about fire prevention.[13]

Denver is not alone in its dependence on forest lands for its water supply. National forest lands are the largest single source of water in the United States and form the headwaters of a large part of the nation's drinking-water supply. In the west, national forests provide proportionately more water because they include the major mountain ranges and headwaters of the principal rivers. All told, more than nine hundred cities rely on national forest watersheds for their public water supply.[14] Decades of aggressive wildfire suppression has resulted in unnaturally high fuel loading within forests. With population growth and climate change projected to increase the frequency and magnitude of wildfires, protection of these watersheds has taken on new urgency. It's also yet another driver for water reuse as part of a diverse set of water supplies.

Straddling the Continental Divide, the majority of Colorado's rivers begin as snowmelt high in the Rocky Mountains. Scenic canyons carved by the runoff provide world-class hiking, fishing, and paddling. The rivers also provide most of Colorado's drinking water and are the lifeblood of the state's farms and ranches.

Colorado may appear water-rich to today's casual mountain visitor. In reality, the state faces a large potential water-supply shortfall within the next few decades. All of its major rivers originate within the state, but Colorado is legally bound to deliver about two-thirds of its streamflow to downstream states to meet river compacts and two Supreme Court decrees. The major rivers on the eastern side—the Arkansas, Rio Grande, and South Platte—have long been over-appropriated. On the Western Slope, the Colorado River and its tributaries are under immense stress. Future demands exceed supply, with climate change making the situation worse. Although separated by more than a thousand miles, San Diego and Denver both face the looming impacts of future shortages on the Colorado River. This "hardest working" river, which supplies water for more than forty million people and nearly 5.5 million acres of farmland across the western United States and Mexico, begins its journey high in Colorado's Rocky Mountain National Park.

Like California, there's a huge geographic mismatch between where people live in Colorado and its water. More than 80 percent of Colorado's population lives east of the Rockies along the Front Range in cities like Denver and Colorado Springs, while almost 80 percent of Colorado's snow and rain falls on the Western Slope. With the population projected to nearly double from 2015 to 2050, the state faces a basic quandary—where will the water come from to meet the needs of its burgeoning population?[15]

Fortunately, the state has a plan. The Colorado Water Plan was released in 2015 following several years of stakeholder involvement and input from over thirty thousand people across the state.[16] The plan provides a basic roadmap, but its goals won't be easy to achieve.

Much of the water to fuel Colorado's urban growth has come by bringing water from the Western Slope through trans-basin diversions that rely on a vast network of storage and conveyance infrastructure. Twenty-four tunnels and ditches move an annual average of a half million acre-feet of water from the Western Slope to the Front Range. This amount is roughly equivalent to the water-supply gap (the difference between existing municipal and industrial supplies and future needs) projected by the Colorado Water Plan.

Denver Water gets half of its water supply from tributaries that flow into the Colorado River. However, the days of largescale dipping into the Western Slope water bounty are coming to an end. The issue came to a head in 2006, when Denver Water filed paperwork for trans-basin diversions to expand Gross Reservoir along the Front Range north of Denver. Planning for the expansion had begun largely in response to vulnerabilities highlighted by the ongoing severe drought and the Hayman Fire. Denver Water gets most of its water from its South System, which had been ravaged by the fire. Its North System, centered around Gross Reservoir, had almost run out of water that same year.[17]

The history of Gross Reservoir is another illustration of how views have changed about Front Range cities tapping Western Slope water. At the dam's dedication in 1955, Denver Water reported that the highlight of the ceremony came when Miss Colorado (and future Miss America) smashed a bottle of Western Slope water against the dam "with dignity and beauty."[18] In the 1980s, expansion of Gross Reservoir had been one of the suggested alternatives to Two Forks Dam. But that was then, and this was now. After meeting with several angry Western Slope entities, Denver mayor John Hickenlooper suggested engaging the services of a mediator.

The process took five years but worked better than anyone could have imagined. In April 2012, Hickenlooper (now governor) announced the Colorado River Cooperative Agreement, a landmark arrangement between Denver Water and local governments, water providers, and ski areas on the Western Slope. It addressed many longstanding disputes and established agreements to cooperate on numerous issues of managing the Colorado River Basin for water supply as well as the aquatic environment. The Western Slope entities consented to expansion of Gross Reservoir. For its part, Denver Water agreed to no further water development from the Colorado River Basin without Western Slope consent. The agreement stresses conservation and reuse by Denver Water.[19]

A second landmark agreement in 2019 gave additional incentives for conservation and reuse. A two-decade drought in the Colorado River Basin finally forced the seven basin states to the table for an agreement that had long eluded them—a drought contingency plan for water sharing in times of shortage. The seven-year agreement is a temporary fix to buy time for stronger steps to address a hotter, drier future.[20]

In addition to trans-basin diversions, farm-to-city transfers along the Front Range are also controversial. Colorado's Water Plan concludes that, without action, the state could lose up to 20 percent of irrigated agricultural lands statewide and nearly 35 percent in the South Platte, its most productive basin.[21] Particularly troublesome are "buy-and-dry" transactions that suck the lifeblood out of agricultural communities. It is widely recognized that additional water sharing between farms and cities is inevitable but should be done in a way that protects the state's agricultural heritage (such as leasing water during dry years and water banking during wet years). While many people conjure up images of Colorado's majestic mountains, farms and ranches also contribute immensely to the state's scenic beauty, open space, and wildlife habitat.

Colorado's Water Plan clearly states that its success will be measured not only by how well the future water-supply-demand gap is closed but also by "the choices we make to close it."[22] The plan calls for increased water conservation to cut consumption by 400,000 acre-feet a year. It also sets a target of reusing 61,000 acre-feet a year by 2050. Reuse is clearly not the sole solution to future water needs, but Front Range utilities have every reason to "push the practical limit" in reusing water.[23]

Interest in water reuse in Colorado goes back decades, with Colorado Springs leading the way. Local water supplies were insufficient for the newly established Air Force Academy, so water rights were purchased from the Western Slope. There was, however, a catch. The newly signed Blue River Decree of 1955 mandated due diligence in reusing trans-basin water to minimize the need for water transfers from the Western Slope.[24] The Blue River Decree was part of a long and complicated litigation over rights to the Blue River, and eventually led to construction of Dillon Reservoir, Denver's largest reservoir. It was perhaps the first major regional agreement in the United States that mandated water reuse by communities.

In 1957, a water-reuse treatment plant was built at the Air Force Academy to irrigate fields, parade grounds, and a golf course. The city of Colorado Springs followed in the 1960s with use of reclaimed water for landscape irrigation. In 1971, Fort Carson began using reclaimed water for golf-course irrigation. The army also recycled the water used to wash its tanks. By 1975, seven Colorado municipalities were practicing water reuse. This was a good start, although California led the way with 138 municipalities reusing water at the time.[25]

Initial state regulations in 2000 focused on landscape irrigation of public areas, but within a few years were expanded to include various commercial and industrial uses. In 2018, the Colorado General Assembly extended allowable uses of reclaimed water to edible food crops, toilet flushing, and industrial hemp cultivation. Legislators said no to a fourth bill that would have allowed use of reclaimed water for the state's booming marijuana cultivation. Cannabis growers feared they would be mandated to use reclaimed water and argued that ample research hadn't yet been conducted on potential health impacts.[26]

DENVER

In the 1970s, Denver Water operated a small pilot plant as a first step in exploring water recycling.[27] More notable was the utility's demonstration plant for direct potable reuse that was operational from 1985 to 1991.[28] As the first direct potable reuse pilot project in the nation, it garnered considerable attention. Delegations of engineers from Europe and the Soviet Union visited the state-of-the-art facility. "There was a sense we were ahead," recalled Myron Nealey, a Denver Water engineer who worked on the project.[29]

From a scientific perspective, the pilot plant was a roaring success. The treated water satisfied EPA drinking-water standards at the time, no adverse health or reproductive effects were detected in rats and mice, and the water was purer than domestic water supplies. Denver Water concluded that potable reuse was a viable option but decided not to move beyond the experimental stage. This decision was not because of pushback from the public but rather came from Denver Water staff concerned about costs, public perception, and regulatory uncertainty. The utility believed that direct potable reuse would someday be needed but decided to focus on nonpotable reuse instead.

A large recycling plant was built at the site of the demonstration project and next door to Denver's Robert W. Hite Treatment Facility, the largest wastewater treatment facility in the Rocky Mountain West. As treated wastewater from the Robert W. Hite facility enters the recycling plant, it flows upward through tanks (called cells), where it undergoes a process known as biologically aerated filtration. The cells are filled with styrene-based media that provide lots of surface area on which microbes that consume organic material can grow. Air is added to the bottom of the cells to provide oxygen for the microorganisms. The tightly packed media also provides filtering. Sticky coagulants are then added to help draw together tiny suspended particles, with the clumping action aided by slow-moving paddle wheels to increase the contact of solids and coagulant. The water then flows through filter beds containing anthracite coal to trap remaining solids. Finally, chemicals are added for disinfection and corrosion control before the recycled water is sent on its way to customers. Recycling operations began in 2004, during the third year of severe drought. The start-up was none too soon.

With a capacity of thirty million gallons per day, a seventy-mile net-work of purple pipes now reaches out to about one hundred customers, serving parks, schools, golf courses, and other operations throughout Denver. Xcel Energy uses about a third of the recycled water in cooling towers at its neighboring Cherokee power plant. With some demand each month, Xcel is considered a prized customer by Denver Water. In contrast, reclaimed water for landscaping is needed only in warm-weather months. This demand-supply mismatch has prompted Denver to push for expanded uses of reclaimed water for purposes beyond irrigation. In a good example of forward thinking, Denver International Airport was constructed with purple pipes for the day that recycled water makes its way out to the airport. Use of recycled water for various uses, from parks to zoos to wildlife refuges, has presented unique challenges in coming up with sensible regulations tied to specific uses.

Once described as "treeless, grassless, and bushless," Denver residents have long had a love affair with green lawns and parks.[30] Even during the Great Depression and drought of the 1930s, one publication called Denver the "City of Beautiful Lawns." Another claimed (with obvious exaggeration) that "Denver has more and greener lawns than any other city of its size in the world."[31] In 1981, to counter this love affair, Denver Water coined the word *xeriscape* (a combination of *landscape* and the Greek word for "dry," *xeros*) to encourage its customers to use plants that are native and adaptable to Colorado's semi-arid climate.[32] Lawns do provide one advantage to water utilities; they offer a cushion during droughts as a relatively painless target for temporary cutbacks.

Denver saves potable water by using recycled wastewater for irrigation to keep the city green, but there's a potential impediment to this practice. The recycled water is higher in salinity than the clear Rocky Mountain water traditionally used. Denver's trees are largely non-native and sensitive to salt. In 2015, controversies arose that the recycled water may be killing trees in city parks, and Denver residents began to question whether the financial savings were worth it.[33]

The utility's technical staff had long been aware of the salinity issue and sold the water at a discount. Ideally, customers would use these savings to mitigate the problem; for example, by "overwatering" to push salts below the root zone and adding gypsum (which releases sodium). Unfortunately, customers pocketed the savings for other purposes.

The issue pitted tree lovers and citizen groups against Denver Water. To everyone's credit, the water utility, parks and recreation, concerned citizens, university experts, and consultants worked together to tackle the issue. Long-term sampling of trees and soils was initiated to better understand the problem, and recommendations were made for management of recycled water sites. Denver Water has funded about a million dollars of improvements to recycled water sites.[34]

The Denver Zoo used recycled water for animal consumption, irrigation, and bathing pools in its "crown jewel" elephant exhibit. The recycled water meets the drinking-water standards of the early 1980s and is definitely cleaner overall than what elephants drink in the wild. It was approved by the zoo's veterinarians, and the animals showed no hesitation in gulping it down, nor adverse health effects. After several years, however, Denver Zoo officials discontinued using recycled water for the elephants to drink—although it is still used for bathing pools and cleaning. The problem was that federal animal-welfare regulations require that the drinking water must meet human-consumption standards, with no allowances being made for the possibility of using high-quality recycled water.[35] It's an example of the need for regulations to catch up with modern-day uses of recycled water.

And then there's Rocky Mountain Arsenal. Located just ten miles northeast of downtown Denver, the arsenal was established as a chemical weapons manufacturing facility after the attack on Pearl Harbor. It later added pesticide factories to the mix. The arsenal caught the attention of Denverites in the 1960s, when underground injection of chemical wastes caused earthquakes that were felt throughout the metropolitan area. In the 1980s, one of the largest environmental cleanups in history began under Superfund and was completed in 2010.

Despite its toxic chemical history, the arsenal was a popular site for bald eagles, leading to its current designation as the Rocky Mountain Arsenal National Wildlife Refuge. In March 2007, the first buffalo in a century to roam the prairie east of Denver were introduced. To maintain the former arsenal as a wildlife refuge, millions of gallons of water are used to fill its lakes and irrigate bison and bird habitats. The arsenal seemed an obvious candidate for using recycled water in place of public supplies of drinking water quality. Surprisingly, given the arsenal's history, the wildlife refuge had to jump through extensive hoops with state and federal authorities before it could switch to using Denver's recycled water.[36]

The most novel use of the recycled water is at the Denver Museum of Nature & Science, where it's used to heat and cool a large section of the museum. This idea works as follows: Because the earth beneath the surface has a relatively constant year-round temperature, the recycled water running through underground pipes is naturally warmed by the earth in the winter and cooled in the summer. In cold weather, a heat pump transfers heat from the water in the pipes to the building. This process is reversed in hot weather. The system reduced the energy required to heat and cool the museum section by 60 percent. No water is lost in the process, so the recycled water can continue toward another use.[37]

In addition to ongoing nonpotable applications, Denver Water continues to explore potable reuse, including a public demonstration project in 2018. While potable reuse appears destined for Denver's future, Aurora, the city's neighbor to the east, is already there.

AURORA

Aurora is Colorado's third largest city. Once mostly White and aging, immigrants from all over the world have transformed the city in a matter of decades to a diverse and relatively young population.[38]

Aurora relied on Denver Water for its water supply for many years. By the early 1950s, however, Denver was having difficulty keeping up with its own population growth and drew a Blue Line beyond which it would no longer grant permits for new water taps. Parts of Aurora were outside the line. For those within, Denver Water would offer contracts for service, but at higher rates and subject to annual mountain snowpack conditions. The writing was on the wall—Aurora needed to develop its own water supply to meet its growing demands.[39]

Alluvial wells were drilled along Cherry Creek (a tributary of the South Platte), and Aurora Water, the city's utility, bought water rights from places such as Last Chance Ditch. In 1968, Aurora's Sand Creek Water Reclamation Facility began providing reclaimed water to irrigate golf courses and parks.[40] Following the tradition of Front Range cities, Aurora also turned to the Western Slope to meet its water demands. When Homestake Reservoir was completed in 1967, Aurora finally achieved its goal of water independence. Western Slope interests and

Denver Water (which wanted in on the deal) had unsuccessfully tried to stop the project but were overruled by the Colorado Supreme Court.[41]

Things turned out differently in the 1980s, when Aurora and Colorado Springs returned to the same area to claim additional water rights left on the table. Known as Homestake II, their plan was to divert water from the Holy Cross Wilderness, south of Vail. Water development in a wilderness area is usually prohibited, but in this case the project had received a special exemption when the wilderness area was created. Aurora and Colorado Springs fully expected the project to move forward, despite opposition from environmentalists and ski resorts.

In a surprise move, county commissioners rejected the permits for the diversion project—a decision upheld by the Colorado Court of Appeals and Colorado Supreme Court. According to the courts, the counties couldn't deny water rights, but they could deny a particular water project by an outside entity. Aurora and Colorado Springs appealed to the U.S. Supreme Court, but the high court declined to hear the case. Occurring at about the same time as the rejection of Two Forks Dam, it was another lesson on the power of mountain communities and environmentalists to kill big water-development projects. Three decades later, the two cities are still trying to develop their water rights in the area.[42]

Aurora Water continued pursuing new water sources, reaching out far and wide into a dozen reservoirs and lakes distributed among the South Platte, Colorado, and Arkansas River basins. The utility managed to keep up with demand until 2002, when the drought hit. By early 2003, Aurora's reservoir levels had dwindled to less than 30 percent of capacity. A blizzard in March 2003 postponed the day of reckoning, but the scare wasn't forgotten. With necessity being the mother of invention, Aurora conceived the Prairie Waters potable-reuse project in 2004. Six years later, the first glass of recycled water was served to the public.

The Prairie Waters Project increased Aurora's water supply by 20 percent and was completed much faster than any new reservoir could have come online. It was also well under budget—a virtual impossibility for a dam project. Prairie Waters treats about ten million gallons per day. "Prairie Waters was huge, not just in terms of volume, but also because it's really helped us advance as a state in accepting potable reused water," emphasizes Laura Belanger, a water-reuse expert with Western Resource Advocates.[43]

Prairie Waters captures water from the South Platte River downstream from Denver and transports it to Aurora for treatment to drinking water standards. It's considered potable reuse because for more than half of the year most of the water in the South Platte comes from the Robert W. Hite wastewater-treatment plant.

Water is drawn from the South Platte by pumping about two dozen wells located approximately three hundred feet from the river. During the seven- to ten-day travel time from the river to the wells, natural processes in the river alluvium remove microbes and chemical contaminants.[44] The process, known as *riverbank filtration* (or simply *bank filtration*), is relatively uncommon in the United States, but many European cities, such as Berlin and Budapest, have used bank filtration from rivers and lakes for over a century as a relatively inexpensive form of pretreatment. Studies in Germany have found that bank filtration is effective in reducing many pharmaceuticals—a major concern when the Prairie Waters Project was designed.

Riverbank filtration is just the first step. Most of the water receives a second underground treatment by recharging spreading basins, where it percolates through sand and gravel for twenty to thirty days before being pumped back to the surface. From there, the water is piped thirty-four miles back to Aurora. Three different pump stations lift the water about one thousand feet on its way to the Peter D. Binney Purification Facility.

At the Binney plant, the water undergoes several treatment steps. Chemical softening reduces nuisance elements—such as iron, manganese, and calcium. Then high-intensity ultraviolet (UV) light, combined with hydrogen peroxide, kills viruses and oxidizes trace organics. This is the same advanced oxidation process that is used in Orange County, California. In the next step, granular filters remove remaining particles and pathogens. In a final step, water passes through giant granulated activated-carbon filters—the same process as the much smaller versions of filters in water pitchers and many home treatment systems. This step is intended to remove any remaining trace organics and improve the taste. Plant operators monitor water quality with an array of tests every four hours, to ensure the reused water meets standards and tastes good enough to deliver to customers. Before it's delivered to homes, the recycled water is mixed with the city's other supplies derived from relatively clean mountain snowmelt.[45]

Not all waters are legally reusable. As in other western states, surface water is allocated in Colorado by the prior appropriation doctrine—first in time, first in right. Under this system, one can't divert water that interferes with more senior rights downstream. As a general rule, reuse cannot decrease natural historical flows in the South Platte.

Fortunately, water from most of Aurora's water sources can be reused. Aurora has the right to recapture and reuse the water that it imports from the Colorado and Arkansas basins. The reusable portion is mostly from indoor uses that return water to the wastewater-treatment plant and from there to the river system, but also includes lawn-irrigation water not consumed (evaporated) after landscape application. Aurora monitors groundwater levels throughout the area to estimate how much lawn-irrigation water is returning to the South Platte and available for reuse. Under Colorado state law, legally reusable water sources can be used over and over to "extinction." The term *extinction* is a bit of a misnomer, because the water doesn't go away.

The key point is that Prairie Waters can treat water multiple times until it's lost from the system, which happens primarily through evapo-transpiration. Aurora Water continues to seek more reusable water and is gradually adding more wells along the South Platte. The pipelines and pumps were designed for fifty million gallons a day, almost five times current use, allowing for plenty of physical capacity for expansion.

Aurora Water does not include reverse osmosis in its treatment train in order to avoid creating brine that must be disposed of. While communities near the coast are able to dump the brine into the ocean (not without its own set of controversies), inland communities must dispose of brine either by land disposal or deep well injection. The brine also ends up wasting some of the precious water that's being recycled. Aurora Water addresses the higher salinity of its Prairie Waters Project by blending the recycled water with other sources. The water will become saltier with time and eventually need to be dealt with in other ways.

The public has been remarkably satisfied with the Prairie Waters Project. There has been no water-quality complaint with water from the Binney plant in the eleven years that it has been operating.[46] The water is expensive, however, costing perhaps ten times as much as water imported from the Homestake Reservoir and other mountain sources.[47] The utility has found a way to reduce these costs through an innovative water-sharing agreement with Denver's South Metro communities.

SOUTH METRO AND A WISE PARTNERSHIP

Compared to Aurora and Denver, water development in the South Metro area took a decidedly different trajectory. From their beginnings, Aurora and Denver aggressively sought renewable water sources to meet the growing needs of their populations. Comparatively late in the game and with limited options for surface water, South Metro communities turned to groundwater from the underlying Denver Basin aquifer system for the vast majority of their water supply. Unlike the alluvial aquifers of the South Platte basin, the Denver Basin aquifers are largely confined by low permeability claystone and shale layers. They receive little recharge from precipitation, and much of the groundwater is essentially nonrenewable.[48]

In the 1990s and early 2000s, Douglas County (south of Denver) was perennially ranked among the fastest growing in the nation. In 1970, the county had a population of just over eight thousand. By 2010, almost three hundred thousand people called Douglas County home.[49] Having staked their future on a nonrenewable resource and with no limits placed on growth, an eventual public outcry and reckoning was inevitable.

In 2003, with drought underway, the *Rocky Mountain News* ran an explosive three-day series, "Running Dry," on the looming water crisis in the South Metro region. The series brought public attention to rapidly declining water levels in Denver Basin aquifers. Groundwater levels in some parts of the aquifer system were falling about thirty feet per year.[50]

According to the banner on the newspaper cover, "Much of Douglas County's well water, once thought abundant for a century, could drop out of reach in 10 to 20 years."[51] Although sensational, the basic message was not a surprise to anyone who had been paying attention. The Denver Basin aquifer system doesn't have the same recognition as other major aquifer systems, such as the High Plains (Ogallala) aquifer, but its long-term sustainability is likewise in jeopardy.

Denver Basin aquifers serve as a water savings account built up over the eons for the South Metro communities, with withdrawals now far exceeding deposits. With negligible recharge, groundwater pumping in the Denver Basin is essentially mining the resource through a planned depletion. Known as the one-hundred-year rule, water from underground sources for a particular property is limited to 1 percent per year of the recoverable groundwater originally underneath that parcel. This law was developed by a group of water engineers and attorneys in the early 1980s (with apparently too few hydrogeologists).

The basic idea was to limit withdrawals to make the savings account last at least one hundred years. Meanwhile, the explosive growth could continue unimpeded. The thinking was that, somehow, a "smooth landing" would take place in transitioning to renewable sources—envisioned largely as surface-water rights purchased from others and transported to the area. State Senator Ken Gordon (D-Denver) described this water policy as "like somebody jumping out of a 90-story building thinking they'll figure something out before they hit the ground."[52]

One hundred years sounds like a long time, but in many cases, South Metro communities are well on their way to depleting their groundwater. There's also a fundamental flaw in assuming that the one-hundred-year rule leads to a one-hundred-year supply. To understand why, we need to look at how water levels in wells respond to pumping in confined aquifers such as those in the Denver basin.

When tapped by a well, groundwater in a confined aquifer rises above the top of the aquifer as a result of the artesian pressure that has built up.[53] As the pressure is reduced due to pumping, water levels can drop precipitously. Water withdrawals from neighboring wells exacerbate the problem. As groundwater levels drop, well yields decline. To pump the same volume of water, wells must be deepened, or additional wells drilled. Meanwhile, water (which is heavy) has to be lifted to the land surface from increasing depths. It can eventually become what's known as "paper water"—it's down there, but getting it to the surface is too expensive.

In 1989, a report by Douglas County's water advisory board questioned whether the county could continue to rely on groundwater. The study recommended requiring that all new developments in population centers use surface water—not a message that developers and politicians wanted to hear. The board members were belittled as "a bunch of anti-development kooks." Meanwhile, wells on the western edge of Douglas County began to go dry.[54]

In 1995, Colorado governor Roy Romer made a dramatic, but unsuccessful, call for Douglas County to put signs in front of new home projects warning buyers that water pumped from the aquifers was unreliable over the long term. The following year, the state engineer's office began placing warning labels on municipal well permits that water supplies may last less than one hundred years. The reports and warnings did little to slow the mining of groundwater. The county growth curve continued unabated.[55]

Even in 2003, when the *Rocky Mountain News* drew attention to the problem, many real estate agents, developers, and elected officials steadfastly assured residents that there was no problem. "We have a 100-year supply, maybe a 500-year supply," asserted a Douglas County commissioner and chairman of the county Water Resource Authority.[56]

In 2004, thirteen water providers formed the South Metro Water Supply Authority (South Metro Water) with a singular focus—to reduce the region's dependence on nonrenewable groundwater. Progress toward this end gained momentum in 2008 when South Metro Water signed an agreement with Aurora Water and Denver Water to explore opportunities for sharing water and infrastructure. They cleverly call this collaboration WISE, an acronym formed from the bureaucratic-sounding "Water Infrastructure and Supply Efficiency." After years of planning and development of critical infrastructure, water deliveries to WISE partners began in 2017.[57]

The regional partnership provides new supply by combining unused capacities in Aurora's Prairie Waters treatment facility with unused water supplies from Denver and Aurora. By sharing treated water supplies when available with South Metro WISE partners, it's a win-win-win for the three parties: Ten South Metro Water members receive significant quantities of new, renewable water supply. Denver Water receives a new backup water supply to use during emergencies or severe drought. Finally, Aurora Water receives additional revenue by making fuller use of its Prairie Waters facility. A less obvious, but important, benefit is a steadier flow of water through the Prairie Waters facility. Treatment plants work best when operated at a constant rate. Biological treatment, in particular, is negatively impacted when turned on and off.

Colorado River users also benefit from the WISE agreement. Increased reuse of water imported from the Colorado River provides a more sustainable supply for Front Range cities without additional Colorado River diversions. As part of the Colorado River Cooperative Agreement, a surcharge on WISE water goes to support river enhancements within the Colorado River basin. Both ranchers and trout benefit. River enhancements run the gamut from stabilizing riverbanks and reviving irrigation channels to creating meandering streams for fish habitat. So WISE is actually a four-way win. Or even five-way, as rural residents of the South Platte and Arkansas River basins are also potential winners in light of less pressure for South Metro utilities to come shopping for their water rights.

Under the WISE agreement, Denver and Aurora must provide a total of at least one hundred thousand acre-feet of treated water to South Metro members every ten years. The amount of water delivered varies from year to year. Aurora and Denver have first dibs on the water and naturally want to use it when they need it the most, which means sending less water to South Metro utilities during dry or high-demand years. In the most extreme case, Aurora and Denver can stop water deliveries for some years. But there are constraints, so they don't dump all the water on South Metro communities in a few wet years.

The allocations are determined through an innovative web portal. Each day, Aurora Water and Denver Water make an offer. Each of the WISE partners has a base allocated amount that, if offered, they have to pay for whether they take it or not (so-called "take or pay"). The WISE water providers can negotiate with other partners to use part or all of their share.

One of the challenges of water from WISE (or other renewable water) is that it may arrive when it's not needed. Having historically depended on groundwater, which can be pumped as needed, South Metro utilities developed very little surface-water storage. Thus, a key part of the WISE partnership focuses on increasing water storage. As one example, Rueter-Hess Reservoir, completed in 2012, was the first major water-storage facility on the Front Range in decades.[58] Storage space in this reservoir is shared by several WISE partners. Using Denver Basin wells for aquifer storage and recovery is also underway. The goal is not to be completely off the Denver Basin aquifer but rather to repurpose the aquifer as a drought supply.

South Metro Water has ambitious goals, aiming to shift 85 percent of the region's water supply to renewable and reuse sources by 2065. To help achieve this goal, South Metro Water has actively promoted conservation and water efficiency, including development of a model landscape and irrigation ordinance. In recent years, per-capita water demand is down by over 30 percent making it one of the lowest in the state, already surpassing the 2050 goal set for the region in the Colorado Water Plan.[59]

The water provided to South Metro Water utilities through the WISE partnership is primarily reusable return flows from its existing supplies (nontributary groundwater is also reusable). This has incentivized South Metro water providers to develop their own local water-reuse capabilities to build on the Prairie Waters Project. The town of Castle Rock is leading the way.

Named for a prominent rectangular butte overlooking the town center, this once-small rural town is now a bedroom community for the Denver metropolitan area, growing more than eight-fold from the early 1990s to 2018.[60] Long reliant on nonrenewable groundwater, Castle Rock has been pivoting to renewable water supplies and potable reuse since the severe drought of the early 2000s.

In 2018, after nine years of planning and investing more than $50 million in infrastructure, Castle Rock began importing WISE water. As a novel way to publicize this milestone, nine humorous short videos on the town's website follow WISE water's journey from Prairie Waters to Castle Rock. Each video ends at a bar where "the Most Hydrated Man in Castle Rock" toasts viewers with a glass of water. "I don't always drink water, but when I do," he assures us, "I prefer Castle Rock water. Stay hydrated my friends!"[61]

Water reuse is not a new idea to Castle Rock. As far back as 1982, the town's plan stated that wastewater "should be completely reused within the community."[62] Some purple pipes were laid, but nonpotable reuse was soon viewed as too expensive. It wasn't until 2019 that the first application came online for irrigating a golf course. Castle Rock's recent venture into potable reuse is taking a much different trajectory.

Beginning in 2006, the town embarked on extensive public outreach for potable reuse that included mailers, support by community leaders, customer surveys, social media, open houses, community events, a town academy, and a website. As a result, potable reuse has been well received by almost everyone. Any skepticism most likely comes from recent residents who haven't been exposed to the town's educational outreach over the years.[63]

Getting to potable reuse was a multistep process. East Plum Creek, a relatively small tributary to the South Platte River, is used as the environmental buffer. In 2013, the Plum Creek Water Purification Facility, initially designed as a conventional drinking-water treatment plant, came online upstream from the wastewater-treatment plant. An off-stream reservoir was built along Plum Creek several miles downstream from the wastewater-treatment plant, along with a pumping station and pipeline to transport the water back from the reservoir to the water-purification facility. After completion of bench-scale and pilot studies, the water-purification facility added advanced-treatment processes. In February 2021, Castle Rock began introducing recycled water into the town's drinking-water supply. Ultimately, the reuse water is expected to constitute about a third of the town's water supply.[64]

Some water is lost in transit, particularly when droughts cause the natural stream to dry up. Looking to eliminate these losses in the future, Castle Rock is positioning itself for possible direct potable reuse. The pipeline has a turnout that would allow connection of the wastewater and water-purification plants, and the treatment processes were selected to meet expected direct-potable-reuse regulations. Castle Rock is clearly on the vanguard of Colorado communities in both indirect and direct potable reuse.

Chapter Seven

Georgia and Virginia Have Water Reuse on Their Minds

Certainly, a city which is only one hundred miles below one of the greatest rainfall areas in the nation will never find itself in the position of a city like Los Angeles.

—Atlanta Mayor William B. Hartsfield (1948)[1]

When it comes to water availability and use, there are two Georgias—the coastal plain in the south and the piedmont/Blue Ridge in the north.[2] The southern half with its agricultural inland and coastal cities depends mainly on groundwater from aquifers composed of sand and limestone layers separated by clays. In contrast, Georgia's urban and industrial northern half relies mostly on surface water. The underlying igneous and metamorphic rocks form low-yielding aquifers. The two regions are divided by the Fall Line, named for the waterfalls and rapids that naturally form as rivers cross from the hard crystalline rocks of the piedmont to the soft sedimentary rocks of the coastal plain.

Water reuse also differs between the two Georgias. In the coastal plain, water reuse is driven mostly by concerns about saltwater intrusion. New permits to pump groundwater in Georgia's coastal counties generally require an evaluation of the feasibility of nonpotable water reuse.[3]

In the piedmont, water reuse is driven by water scarcity in the face of a growing population. This is particularly true in metro Atlanta, where roughly half of the state's residents live. The Atlanta area focuses on potable reuse. Nonpotable use for new landscape irrigation is discouraged.[4]

Metro Atlanta's principal source of water supply is Lake Lanier, a large reservoir created by the completion of Buford Dam on the Chattahoochee River in 1956. The U.S. Army Corps of Engineers (Corps) operates the dam for multiple competing purposes, including hydropower, flood control, recreation, and Atlanta's water supply.

Lake Lanier is by far the largest reservoir in the Apalachicola-Chattahoochee-Flint (ACF) River Basin. Georgia's downstream neighbors, Alabama and Florida, want their fair share of this water bounty. Three decades of water wars with Alabama and Florida have been fought over the ACF. A primary issue is whether or not metro Atlanta has a right to water from Lake Lanier, and if so, how much. The seeds of this conflict were planted over seventy years ago when the dam was first contemplated.

In the late 1940s, Atlanta Mayor William B. Hartsfield lobbied intensively for Buford Dam, shuttling back and forth to Washington, often with key Georgia business leaders in tow.[5] Hartsfield was an astute politician who served as Atlanta's mayor for a quarter century (1937–1962). He is best known for his key role in turning the site of a dirt racetrack into Hartsfield-Jackson Atlanta International Airport—the world's busiest passenger airport. But his fine-tuned political radar failed him at a critical point on Buford Dam. In response to congressional requests for Atlanta to help pay for the costs of dam construction, Hartsfield balked, firing off a missive that included the quote at the beginning of this chapter:

In 1951, three years after Hartsfield's note, first-term congressman Gerald Ford of Michigan speculated about the long-term consequences of Atlanta's refusal to help fund the dam: "Is it not conceivable in the future, though," Ford asked, "when this particular project is completed, that the City of Atlanta will make demands on the Corps because of the needs of the community, when at the same time it will be for the best interests of the overall picture . . . to retain water in the reservoir?"[6]

The future president's warning about Atlanta playing second fiddle in competition for the reservoir's water proved to be prescient. In 1990, Alabama and Florida filed federal lawsuits to stop metro Atlanta from taking more water from Lake Lanier. The ensuing tri-state water war is well known, but a quick review of a few milestones illustrates what was at stake.

In 1997, the three states reached an agreement known as the ACF River Basin Compact. This was basically an *agreement to agree* on an allocation formula. It never happened. The compact died in 2003, and the three states went back to court.

In 2007, a drought was pushing Atlanta almost to the point of disaster. Georgia had virtually no leverage over the corps' operations of the reservoir, which was releasing water for downstream uses in Florida and Alabama (and for endangered species). Georgia Governor Sonny Perdue famously prayed for rain.[7]

In 2009, Judge Paul Magnuson issued a ruling declaring that water supply is not an authorized purpose of Lake Lanier, and imposed, in his words, a "draconian" injunction that would have cut metro Atlanta's water supply in half. Magnuson gave Georgia three years to obtain congressional approval for additional authorization.[8] He cited Hartsfield's 1948 memo as support for his position.[9] The ruling sent shock waves across Georgia. Fortunately for the state, a federal appellate court overturned Judge Magnuson's decision in 2011.

In 2013, Florida sued Georgia in the U.S. Supreme Court, asking for an "equitable apportionment" of the waters of the ACF River Basin that would restrict Georgia's water use to 1992 levels. Florida claims that Georgia's water use has harmed its Apalachicola Bay oyster fishery and caused its collapse in 2012. Georgia asserts that its water use has only a minor impact on the flow in the Apalachicola River at the state line, and that the collapse of the oyster fishery is the result of environmental factors and mismanagement by the state of Florida. In 2021, the U.S. Supreme Court ruled unanimously in favor of Georgia, although the high court also emphasized that Georgia has an obligation to help conserve water in the basin.[10]

In the face of multiple droughts, a fast-growing population, and decades of interstate litigation, metro Atlanta had plenty of incentive to develop potable reuse. A suburban county northeast of the city of Atlanta would take the lead.

In the 1980s and 1990s, Gwinnett County was among the fastest-growing counties in the United States—it even held the number-one spot for several years. The county relies exclusively on Lake Lanier for its water supply. It has virtually no other local water sources, and by state law, the county cannot pipe water from basins in Georgia that are outside the metro Atlanta regional water district.[11]

In 1993, when Wayne Hill was elected chairman of the Gwinnett County Board of Commissioners, water and wastewater were among the top issues in meeting the needs of the county's exploding population. But there was a problem—land that the county had bought for a wastewater-treatment plant was in the wrong basin. Gwinnett County sits astride the Eastern Continental Divide, with half the county draining to the Atlantic Ocean and the other half draining to the Gulf of Mexico via the Chattahoochee River. The facility had to be built in the Chattahoochee River basin and the water returned to that basin.

As an amateur pilot who liked to fly in his spare time, Hill observed a large open area with just a few houses sandwiched between interstates 85 and 985. The land was in the right basin and seemed the perfect spot for a wastewater-treatment facility. Engineers balked at first, saying the land was too high in the basin, but eventually plans proceeded for a large wastewater treatment plant.[12]

Hill envisioned a potable-reuse project using surface-water augmentation similar to the Occoquan Reservoir in northern Virginia. He would periodically take officials to visit the Occoquan plant to see firsthand its successful operation as a model for Gwinnett County. Initially, Hill found it hard to convince people of the merits of potable reuse. Even his brother was against it. But the project was nowhere near as contentious as the San Diego potable-reuse project that was going into a tailspin on the other side of the country.

As the project proceeded, it earned high marks for giving responsibility to citizen groups. A citizen advisory board controlled its own $50,000 budget for technical reviews, sampling, and other activities. As an example of its influence, the board was instrumental in obtaining a county resolution that assured automatic annual increases in labor costs for retraining at the plant. As one board member put it, "We have a highly qualified group running the plant, but if they don't continue training . . . they could become complacent."[13]

In 2000, the advanced wastewater-treatment plant came online with a capacity of 20 million gallons per day. The plant's effluent was discharged to the Chattahoochee River downstream of Lake Lanier, and so was not available for reuse by the county. However, even before the plant was completed, it had become clear that the plant's capacity would have to be expanded to meet the needs of the rapidly growing population.

The same year that the plant came online, the Georgia Environmental Protection Division (EPD) granted a permit for an additional 40 million gallons per day. This treated wastewater would be discharged directly into Lake Lanier, thereby contributing to the county's water supply through indirect potable reuse. As expansion of the treatment plant got underway, a group of lakeside homeowners and businesses, known as the Lake Lanier Association, sued the Georgia EPD. Their lawsuit was not about the squeamishness of drinking treated sewage. Rather, the lakeside group was concerned about the effects of phosphorus on algal growth in the lake.[14]

In 2005, after five years of drawn-out battle, the Georgia Supreme Court ruled in favor of the Lake Lanier Association. To satisfy the court, the utility agreed to very low levels of phosphorus in the effluent discharged to Lake Lanier—the strictest in Georgia, and one of the strictest of any wastewater-treatment facility in the southeastern United States.

In 2006, when the Georgia EPD approved the permit, the Lake Lanier Association was fully onboard. With their concerns about phosphorus addressed, lakeside residents now saw the benefits—the treatment plant would add more water to Lake Lanier, and a full lake is good for everyone. "We applaud them," said Val Perry, the association's executive vice president. "We think Gwinnett acted in outstanding good faith. This was admirable. They did a great job."[15]

With the plant expansion complete, a pipeline still needed to be built to transport the treated wastewater to Lake Lanier. Finally, on May 4, 2010, Hill's successor clicked a computer mouse, a valve opened, and treated wastewater began pouring into Lake Lanier, the county's sole source of drinking-water supply for its now more than nine hundred thousand residents.[16]

The wastewater-treatment facility is named the F. Wayne Hill Water Resources Center, in honor of the person most credited with making it happen. Hill is quick to say that many people are responsible for its creation. The advanced wastewater-treatment processes include ultrafiltration, pre-ozonation, biologically active carbon filtration, and post ozone disinfection to produce high quality water that is returned to Lake Lanier.[17] It's one of the largest advanced wastewater-treatment facilities in the world.

The Hill plant is permitted to pump up to forty million gallons of advanced-treated wastewater each day into Lake Lanier, and another twenty million gallons into the Chattahoochee River. Not surprisingly, the county prefers to put the water into the lake, because any water pumped into the river below the dam will immediately be out of reach by the county as it flows downstream toward the Gulf of Mexico. Gwinnett County currently treats about thirty-five to forty million gallons on most days, and so most of the treated water goes to Lake Lanier. This accounts for more than half the water that the county withdraws from Lake Lanier each day.[18]

The six-foot-diameter pipe that carries the treated wastewater extends a little over a mile into the lake and more than 100 feet below its surface. The pipe's low release point and cooling of the water before reaching the lake help keep the recycled water toward the lake bottom, away from sunlight, so less algae will be formed.[19]

Two drinking-water treatment plants withdraw water from Lake Lanier and utilize ozone biofiltration to produce high-quality drinking water. Interestingly, the Lake Lanier Association had pushed for the outfall for the treated wastewater to be close to the county's drinking-water intake so that it would pick up as much of the treated wastewater as possible.[20] One of the intakes is about a mile from the outfall. This proximity has not been a problem.

The plant pioneered a treatment train that doesn't require the energy-intensive process of reverse osmosis and avoids inland disposal of RO concentrate. The process doesn't remove salts from the wastewater, but unlike in the western states, salt is not a major water-quality issue in northern Georgia.

Chemicals added to help meet the low phosphorus requirement caused deposits to build up on the inside of the plant's pipes. To circumvent this problem, a nutrient-recovery system was installed that removes 85 percent of the phosphorus before it can accumulate on the pipes. The end-product is a slow-release pelletized fertilizer sold to agriculture, turf, and horticulture markets. The plant produces more than a ton of this fertilizer each day, turning a contaminant into a valuable resource.[21]

Looking to the future, the county has been experimenting with direct potable reuse (DPR) using a blend of advanced-treated wastewater and lake water. A pilot project demonstrated that a blend of 15 percent effluent and 85 percent lake water met all drinking-water standards for regulated contaminants, as well as being below action levels for numerous contaminants of emerging concern. Higher blends are possible with minor modifications to the plant processes.

The DPR pilot project also provided a surprise benefit. Lakes often turn over in the fall as a result of temperature differences between the lake surface and deeper zones. Lake Lanier is no exception. The turnover causes increased turbidity (cloudiness from particles suspended in the water) and stirs up organics in the water, requiring more vigorous treatment. A fifty-fifty blend of treated wastewater and lake water did not have these same challenges.

In the 1990s, Gwinnett County had been facing a future drinking-water shortage. Today, the county obtains half its drinking water by recycling wastewater drawn from residences throughout much of the county. The wastewater-treatment facility is one of the world's largest potable-reuse projects using surface-water augmentation. The facility has never had a water-quality violation.[22] In 2018, the Gwinnett County Department of Water Resources won the Excellence in Environmental Engineering and Science Grand Prize for Research for its DPR pilot.[23]

Wayne Hill made other contributions during his twelve-year tenure as commissioner. Two of his last acts were signing the contract and helping to break ground for the Gwinnett Environmental & Heritage Center to be built on the treatment plant campus to educate children. In 2018, at the unveiling of a bronze sculpture of Wayne Hill and three children, Hill noted, "This place is probably more important to me than the wastewater facility simply because of all of the education for the kids that we do."[24]

Gwinnett County continues to look for other opportunities for innovation. In 2019, the county broke ground on "the Water Tower." Sharing the same campus as the Hill plant, the Water Tower has ambitious plans to bring together public, private, and nonprofit entities in a global water-innovation hub. "Our vision is to become no less than a thriving ecosystem of water innovation fueled by imagination, informed by research and powered by pioneers," says CEO Melissa Meeker.[25] The idea for the name dates back to the early 1970s, when a pair of water towers welcomed people to Gwinnett County, unabashedly declaring, "GWINNETT IS GREAT," and "SUCCESS LIVES HERE."

The facility will merge the practical with the innovative through applied research, technology innovation, workforce development, and public engagement. Access to multiple streams from the county's water and wastewater-treatment plants will allow for real-world testing and applied research. Participants will have access to leading industry experts and resources to design, test, and validate the latest technologies. Active recruitment, internship, and apprenticeships will provide a career pipeline for tomorrow's workforce—a widely recognized need with today's ageing workforce and changing technologies.

Clayton County, home to the Atlanta international airport, is another pioneer in potable reuse. While the airport gets its water from the city of Atlanta, the rest of this densely populated county depends on limited local water resources. To meet these challenges, Clayton County has become a national leader in using constructed wetlands for indirect potable reuse.

The story begins in the mid-1980s, when Clayton County began augmenting one of its reservoirs by using sprinklers to apply treated wastewater to forestland adjacent to a water-supply reservoir. After passing through the soil, the reclaimed water flowed into the reservoir. This simple treatment minimized the impact of wastewater discharges on stream quality, as well as returned some of the water to a county reservoir.[26]

Beginning around 2000, as water demands expanded, the land-application system was replaced by a series of constructed wetlands. The wetlands consist of interconnected, shallow ponds filled with native vegetation. Natural processes remove pollutants remaining in the tertiary-treated wastewater as it travels through the wetlands over a period of one to two years on its way to water-supply reservoirs.

The wetlands require less land, less energy, and less maintenance than the land-application system. They have also saved money. Additional benefits include habitat for birds and recreational and educational opportunities, including a popular wetlands center.

The 2007 drought tested the ability of the reuse system to reduce the vulnerability to droughts. While many utilities in north Georgia were in danger of running out of water (including record low levels in Lake Lanier), the county water authority maintained an ample water supply throughout the crisis. "It's raining every day in Clayton County," the utility declared.[27]

VIRGINIA

Ted Henifin, general manager of the Hampton Roads Sanitation District, doesn't remember exactly when, but sometime in the early 2010s, he and his staff had a bold idea on how to simultaneously address several environmental problems vexing the lower Chesapeake Bay region.[28] The Hampton Roads Sanitation District (HRSD) provides wastewater treatment for eighteen cities and counties, covering 1.7 million people in eastern Virginia. Instead of discharging the treated wastewater into Chesapeake Bay tributaries, Henifin and his team's idea was to put it through additional advanced water treatment and use it to replenish the Virginia Coastal Plain aquifer system. The plan would eliminate 90 percent or more of the utility's wastewater discharges to the Elizabeth, James, and York Rivers.

The Chesapeake Bay, the nation's largest and most productive estuary, was an obvious beneficiary of this plan. For decades, the bay has been imperiled by overloading of nitrogen and phosphorus, creating a domino effect. The excess nutrients stimulate algal blooms, which decompose, creating large areas of low dissolved-oxygen concentration that kills aquatic life. The algal blooms also block sunlight needed by submerged grasses. When those grasses die, they remove an important food for waterfowl, and shelter for crabs and young fish. To restore the Chesapeake Bay, the U.S. Environmental Protection Agency established a multi-state pollution diet. Officially known as a Total Maximum Daily Load (TMDL), the diet requires large reductions in nitrogen, phosphorus, and sediment to the bay by 2025.

In addition to the Chesapeake Bay, the plan addressed the regional aquifer system, which badly needs a restoration program of its own. While the treated water discharged to local waterways has no beneficial use, it could prove invaluable in restoring the groundwater resource. In the early 1900s, many wells drilled in eastern Virginia were artesian, with natural pressures causing water geysers as high as thirty feet. These artesian conditions are long gone. Over the past century, groundwater levels have dropped as much as two hundred feet, causing a smorgasbord of effects—decreased well yields, increased pumping costs, land subsidence, and vulnerability to saltwater intrusion from the Atlantic Ocean. In 2017, the state of Virginia cut back withdrawal permits for the fourteen largest groundwater users. Two of these users had to reduce their pumping; the others now have less room to expand.

Modeling studies by HRSD consultants showed that the planned injection of treated water into the aquifer would substantially raise water levels regionwide within fifty years, with significant recovery in some areas as early as ten years. While water levels in the confined aquifer respond relatively quickly to repressurization, the time of travel is slower, taking about two hundred years for the injected water to travel a mile.[29]

The rising water levels would help address another problem. The aquifer system is a layered sequence of sand and gravel aquifers separated by silt and clay confining beds. As groundwater pumping reduces aquifer pressure, the clay layers slowly compact, causing land subsidence. The sinking land, combined with rising seas, results in the highest rates of relative sea-level rise on the Atlantic Coast.[30] There's a lot at stake for the region's low-lying coastal urban areas and sensitive ecosystems. The Naval Station in Norfolk, the largest naval base in the world, is among the most vulnerable military bases to climate change.[31]

To help restore the Chesapeake Bay and other water courses, the HRSD and localities in the Hampton Roads area are under a federal consent decree to reduce sewer-system overflows during wet weather.[32] The price tag could be about $2.4 billion for HRSD and a couple billion more for the localities served by the wastewater utility. In this light, $1 billion for HRSD's alternative plan seems like a bargain, particularly since it would have a larger impact on reducing nutrients to the bay than curtailing sewer overflows would have. The plan can produce federal pollution credits that the localities can use to offset some of their required expensive improvements to reduce sewer overflows. Some stormwater improvements would still be needed—for example, to reduce bacterial pollution that causes closure of beaches and shellfish harvest areas.

In summary, the HRSD's concept, known as the Sustainable Water Initiative for Tomorrow (SWIFT), would concurrently reduce nutrient discharges to the Chesapeake Bay and provide a more sustainable supply of groundwater. By drastically reducing nitrogen and phosphorus content in its treated wastewater, SWIFT will spare the HRSD and localities billions of dollars for stormwater retrofits (at least in the short term). It also would eliminate uncertainty about the possibility of stricter regulations in the future on HRSD's nutrient discharges to the bay by proactively reducing them. Avoiding future saltwater intrusion and slowing the rate of land subsidence and relative sea-level rise are an extra bonus. The HRSD's goal is to recharge about one hundred million gallons per day, by around 2032, from four or five of the sanitation district's treatment facilities.[33]

The HRSD has several advantages in this endeavor. It's governed by an apolitical governor-appointed commission, covers a broad area that encompasses multiple municipalities, and directly bills its customers. These attributes give it a certain independence as well as regionwide outlook.[34] Nonetheless, SWIFT comes with both technical and public-relations challenges. The utility has taken a proactive approach to both.

After several years of developing the concept, the SWIFT Research Center was established at its treatment plant in Suffolk, Virginia. The center includes 1 million gallons per day of advanced water treatment, a recharge well, monitoring wells, a public outreach and education center, and research facilities.

A key technical challenge is to match the chemistry of the injected water as closely as possible with that of the natural groundwater to avoid well clogging and other complications. HRSD builds on the experiences of Chesapeake, Virginia—one of the localities that it serves. For three decades, Chesapeake has successfully operated a well for recharge and recovery of treated drinking water using the same aquifer targeted by SWIFT. Initially, well injection mobilized manganese present in the aquifer matrix, causing discoloration of recovered water, but the problem was solved through pH adjustment.[35]

The SWIFT Research Center explored both a membrane approach using reverse osmosis (similar to Orange County, California) and a carbon-based approach that uses biologically active filtration (like Gwinnett County) and granular activated carbon. The carbon-based approach was found to be much more compatible with the native groundwater and aquifer minerology, eliminating the need for significant post-treatment additives. Removal of salt is a common benefit of RO, but in this case, salt would have to be added back to make the water compatible with the aquifer. The carbon-based approach also saves money and energy.

Building on the lessons of pioneers in potable reuse, the HRSD diligently worked from the outset to get stakeholders, politicians, and the public on board with the idea. They moved quickly to meet with political leaders from the governor to local politicians, explaining how SWIFT is a solution to multiple local water challenges. Utility leaders personally gave numerous briefings and listened to stakeholder questions and concerns.[36]

SWIFT is still in the early stages, but the effort has not gone unnoticed. In 2018, the US Water Alliance recognized the HRSD with the prestigious U.S. Water Prize.[37]

Chapter Eight

Florida

The Long Road from Purple Pipes to Potable Reuse

From what I have observed, I should think Florida was nine-tenths
water, and the other tenth swamp.

—A disillusioned newcomer in 1872[1]

David W. York, Florida's Water Reuse Coordinator from 1980 to
2007, often spoke of the three eras of Florida water reuse.[2] The period
prior to the mid-1980s was the "Dark Ages"—a time of limited reuse
activity and very little institutional framework. It was also the "Age of
Disposal" to surface water, ocean outfalls, and deep injection wells.
Rules governing reuse were limited to spray irrigation and other land
application systems. Public access areas (from parks and golf courses
to residential lawns) were addressed in only one paragraph. Little effort
was made to encourage or facilitate other promising forms of water
reuse. The term *reuse* did not even appear in Florida's rules during this
period.[3]

Florida's Dark Ages of water reuse were not entirely devoid of progress. During this period, the city of St. Petersburg developed the first
large-scale nonpotable water reuse system in the United States and one
of the largest reuse systems in the world.

St. Petersburg sits on a peninsula on the west coast of Florida,
between the Gulf of Mexico and Tampa Bay—the state's largest open-
water estuary. Popular for sport and recreation, the bay also supports
one of the world's most productive natural systems. As the city grew,

Tampa Bay's water quality and abundant marine life were deteriorating, with sewage discharges threatening to destroy this ecological treasure. The city's water supply was likewise challenged by population growth and saltwater intrusion.

In 1972, the state legislature passed the Wilson-Grizzle Act, wherein all wastewater utilities were required to either stop discharging into Tampa Bay or install advanced treatment systems to meet nutrient-reduction requirements. In an effort to simultaneously save the bay and supplement the city's water supply, as well as avoid expensive treatment plant upgrades, St. Petersburg merged its sewer and water departments and in 1977 began an extensive water-reuse program. Excess wastewater that is not used is injected through wells into deep groundwater that's too salty for use as a water supply.

Initially, recycled water was provided only to large users, such as golf courses, parks, and schools. It wasn't long before the system was expanded for residential lawn watering to neighborhoods where a sufficient number of homeowners petitioned to connect to the dual distribution system. Residents had to pay for the hookup but benefited from cheaper nutrient-rich water for their lawns. This reclaimed water is not permitted for indoor uses, sprinkling on edible crops, use in pools, or washing cars, boats, or driveways. By 2009, recycled water was meeting about 40 percent of St. Petersburg's total water demand.[4] Today the city has more than ten thousand water-reuse customers connected by almost three hundred miles of pipelines.[5]

There are limitations to this dual distribution approach. It's hard to get people to conserve, because the reclaimed water is not metered, and customers are charged a fixed monthly rate.[6] The reclaimed water also is not subject to the restrictions placed on potable water use during droughts. Restrictions only kick in when the system experiences low pressure due to demand exceeding supply. It's also far from an unlimited resource. It takes six wastewater customers to produce enough irrigation water for one residence.[7]

In the program's early days, people complained that the reclaimed water was harming their plants. In response, the city funded Project Greenleaf. The study team set up experimental plots and monitored about two hundred ornamental plants at randomly selected residences. The study identified fifteen tree species for which reclaimed water should not be sprayed directly onto the young leaves of saplings (use drip irrigation instead). The water should not be used *at all* on azaleas and Chinese privet, which were found to have extremely low salt tolerances. It was also important to manage chloride concentrations in the reclaimed water for many plants. The complaints died down.[8]

The second era of water reuse was the "Age of Expansion," which took place from the mid-1980s to the early 2000s. This chapter in Florida's water-reuse history began with another landmark project. In 1979, a lawsuit by a citizens group against the city of Orlando and Orange County sought to stop two wastewater-treatment plants from discharging their effluent into Shingle Creek—the northernmost headwaters of the Everglades watershed. The effluent was degrading a downstream lake and its fish habitat.[9]

The court ruled that the wastewater discharges must cease by March 1988. The city and county had to find another way to get rid of their treated wastewater—a challenge compounded by a growing population. They chose to use advanced secondary-treated wastewater for citrus irrigation and aquifer recharge. Citrus growers were initially resistant to the idea but got onboard after research showed that the reclaimed water would be beneficial to their crops. As further enticement, a weekly quota of reclaimed water would be provided to the growers at no cost for the first twenty years.[10]

This cooperative project among the city, county, and agricultural community became known as Water Conserv II.[11] It's one of the largest water-reuse projects of its kind in the world, combining agricultural irrigation with aquifer recharge. It was also the first project in Florida permitted by the state to irrigate crops produced for human consumption with reclaimed water. The system has diversified with the addition of golf courses and residential irrigation. Excess flows not needed for irrigation are diverted to rapid infiltration basins to recharge the Floridan aquifer, the state's primary drinking-water source. The rapid infiltration basins (RIBs) are just over a football field long and about 150 feet wide, built along a natural sand ridge.[12]

The project began operation in December 1986, well ahead of the court-mandated deadline. Agricultural and commercial customers use about 60 percent of the reclaimed water, with the remainder going to the RIBs. A computerized system that forecasts the impact on the groundwater system is used to operate the RIBs.

Water Conserv II turned a liability (wastewater that was contaminating environmentally sensitive surface waters) into an asset (reclaimed water for beneficial use). It reduces the need to pump from the Floridan aquifer for irrigation, while also replenishing the aquifer through the RIBs. The citrus growers get a dependable long-term source of irrigation water that is not subject to water restrictions during droughts. The RIBs also serve as preserves for plants and animals.

During the Age of Expansion, Florida established formal rules and guidelines for water reuse. Many utilities implemented water-reuse programs, and the state's reuse business was booming. But there was an underlying problem—giveaway programs and low flat rates encouraged overuse. With plenty of reclaimed water to go around, little attention was given to its inefficient use until the "drought of the century" brought the problem to light in 2000. Many reuse systems ran short of reclaimed water, angering customers who had been promised an unlimited, drought-proof supply.

In 2003, a multiagency committee proposed strategies to move Florida toward a third era—an "Age of Enlightenment" in water reuse. The plan emphasized increased efficiency in the use of reclaimed water, along with ambitious goals for water reuse. By 2020, 65 percent of all domestic wastewater statewide would be reclaimed and reused for beneficial purposes. Groundwater recharge and indirect potable-reuse projects would become common practice. Sewer mining—small decentralized treatment plants tapping into the sewer system—also would become common practice, enabling more effective use of reclaimed water. Use of ocean outfalls, surface-water discharges, and deep injection wells for wastewater disposal would be largely limited to facilities that serve as backups to water-reuse facilities.[13]

These ambitious goals were not met, but they set the stage for considerable progress on multiple fronts. Florida maintains a careful accounting of its water reuse, and the numbers are impressive. In 2020, 413 wastewater-treatment facilities provided about 884 million gallons per day of reclaimed water for beneficial uses. Over half the water was used for landscape irrigation, including at 442,277 residences, 489 golf courses, 1,005 parks, and 384 schools. The rest was used for industrial purposes, groundwater recharge, agricultural irrigation, and hydrating wetlands.[14]

Florida averages more than fifty inches of rain annually. Much of the state is basically former swampland that's just above sea level. So how did such a water-rich state become the nation's number-one user of recycled water?

The answer is severalfold. Florida is not wet year-round; the rainfall is concentrated during the months of June through September. The state is also susceptible to severe droughts. Florida's three major rivers—the Suwanee, Apalachicola, and St. Johns—are all in the northern part of the state. The population resides mostly in peninsular Florida, where streams tend to be small, slow moving, and warm year-round. They flow into sensitive lakes or coastal waters that are prone to excessive growth of algae, water hyacinths, and other nuisance aquatic weeds. In addition to their environmental impacts, harmful algal blooms are a bane to the state's tourist industry.

Florida spent most of its early days getting rid of water. Environmental journalist Cynthia Barnett notes the irony, "A century ago Floridians thought their biggest problem was too much water where people wanted to settle. Now, our biggest problem is that we do not have enough water where people want to settle."[15]

Wetlands once covered more than half of Florida.[16] The Everglades with its extensive sawgrass marshes is the best-known example, but wetlands are scattered throughout the state. These swamps and marshes were considered nuisances standing in the way of land development and agricultural production. The general view was that they were filled with poisonous snakes and swarms of mosquitos and served as a breeding ground for malaria and other diseases.[17] So-called ditch-and-drain laws were the state's first water laws. The land could then be developed and sold to northerners, who were attracted by advertising campaigns promoting the climate.

Today, it is widely recognized that wetlands provide invaluable habitat for waterfowl, fish, and other wildlife. They reduce flood damages by retaining overflows in backwater ponds and depressions—a particularly useful feature along Florida's hurricane-prone coast. Wetlands also provide water-quality benefits by removing nutrients and other contaminants from water flowing through them. Although now protected by state statute, only about half of the original wetlands remain.

With limited surface water, groundwater provides drinking water to more than 90 percent of Florida's population.[18] This vast underground water system is a product of the state's geological history. For eons, a large, shallow sea covered all of what is now Florida. Gradually, over millions of years, shells of marine creatures accumulated into thick beds of carbonate rocks that form the Floridan aquifer (officially, the Floridan aquifer system). The soluble limestones and dolomites are sculpted by dissolution and weathering into a distinct landform known as karst. As water percolated through the relatively easily dissolved carbonate rocks, it created openings ranging from solution-widened cracks to large caverns.

The Floridan aquifer underlies all of Florida, southern Georgia, and small parts of South Carolina and Alabama. It's one of the world's most productive aquifers, but extensive pumping has caused a variety of impacts that limit future development of the aquifer. (In southern Florida, where the Floridan aquifer is deep and salty, other aquifers, such as the Biscayne aquifer in the Miami–Palm Beach area, are important sources of groundwater.)

The Floridan aquifer is renowned for its more than seven hundred springs. Eight billion gallons of freshwater bubble out each day—more than any similar-size area on earth.[19] The clear, azure waters of Florida's springs are magnets for wildlife, residents, and tourists alike. Nearly everyone who visits them is astounded by their beauty. Recreational opportunities abound, with swimming, snorkeling, diving, and canoeing among the most popular activities. The springs maintain temperatures of about seventy degrees Fahrenheit year-round, and so are great places to cool down during Florida's hot, humid summers.[20]

The springs provide "windows" into the Floridan aquifer to measure its health. But in peering into these windows today, many people do not like what they see—algal blooms caused by seepage of nutrients from farms, urban areas, and septic tanks, along with decreased spring discharge caused by groundwater pumping. Consider an example.

Located just outside the city of Ocala in north-central Florida, the iconic Silver Springs is considered Florida's first bona-fide tourist attraction. The glass-bottom boat was invented here in the late 1870s.[21] Today, the springs face dual threats from groundwater pumping and nutrients. With support from the St. Johns River Water Management District and the Florida Department of Environmental Protection, the city recently built the Ocala Wetland Recharge Park at an abandoned nine-hole golf course. Reclaimed wastewater and treated stormwater pass through a series of wetlands, removing most of the nitrogen and phosphorus and recharging the aquifer. Walking trails and wildlife overlooks provide a pastoral environment for visitors. Previously, the city's excess reclaimed water was conveyed to spray fields for disposal, providing no ecological or aesthetic value and limited nutrient reduction.[22]

Groundwater withdrawals and nutrients also affect lakes and streams throughout the state. Near the coast, where the majority of the population lives, groundwater supplies are vulnerable to saltwater intrusion. On top of these challenges, groundwater pumping has contributed to sinkhole development, particularly in west-central Florida. Sinkholes are natural features in Florida, but pumping can exacerbate their formation. Along with the effects on buildings and other structures, sinkholes can be a source of contamination. In some rural areas, they're known as "go-away holes" for disposal.

These water challenges have motivated the state to become a leader in water recycling. Now reusing about half of its domestic wastewater, the state has long led the nation in the amount of water it reuses. The extent of reuse, however, varies considerably around the state. Water reuse is close to 100 percent in parts of central Florida, 30–50 percent in the Palm Beach area, and 4–7 percent in the Miami area. To date, Florida's water-reuse efforts have focused on nonpotable applications. For a closer look at the intricacies of Florida's water reuse, let's return to the Tampa Bay region.[23]

Pinellas County is a peninsula nearly surrounded by the saltwater of Tampa Bay and the Gulf of Mexico. The county has the highest population density in Florida, with St. Petersburg at its southern tip. As saltwater intrusion made groundwater unsuitable for public supply, the city and county looked to the east and north in rural Hillsborough and Pasco Counties for their water salvation. The first wellfield and pipeline were completed in 1930.[24]

By the 1960s, as Pinellas County and St. Petersburg were aggressively buying land for the purpose of drilling wells, residents living near these wellfields began to notice changes to the natural landscape. Wetlands were vanishing, and lake levels were dropping. Trees died, and wells began to run dry. Sinkholes were developing, damaging foundations, walls, and ceilings of homes.[25] St. Petersburg and Pinellas County leaders were unrepentant, blaming the problems on lack of rainfall. Once it started raining, they assured everyone, things would change.

In the early 1970s, Pasco, Hillsborough, and Hernando Counties successfully lobbied for legislation to block further water development by municipalities outside their jurisdiction. Among the results, this legislation stopped St. Petersburg from tapping into Weeki Wachee Springs—at the time, the deepest known freshwater cave system in the United States and a popular tourist attraction.[26]

In 1974, pressure to find a solution to the region's water woes led to the creation of a regional water-supply authority through a five-party agreement among Hillsborough, Pasco, and Pinellas counties and the cities of Tampa and St. Petersburg. The idea was that the cities and counties would cooperate to develop new water supplies through the regional authority. Unfortunately, they couldn't agree on the new projects to be developed.

In the late 1970s, development of new wellfields allowed pumping cutbacks at the older wellfields, helping the environment to recover somewhat in these areas. But the first round of conflict and resolution was just a preview of what was to come.[27] Environmental impacts due to pumping continued to become more widespread. Big Fish Lake, famous for its large bass, was once thirty feet deep and covered nearly three hundred acres. It went dry around 1990.[28] Others watched as their beautiful lakefront property became a mudflat. People pleaded for the state to step in and help.

Florida assigns responsibility for issuing permits for water supply to five water-management districts. The Tampa Bay region is under the purview of the Southwest Florida Water Management District. After a period of denial and with negotiations going nowhere, the water-management district issued an emergency order to stop the ever-increasing pumping. The order met with stiff resistance. St. Petersburg, Pinellas County, and the regional water-supply authority contended that since the permits were issued, they had every right to pump the water. The water-management district, residents and leaders in Hillsborough and Pasco County, and environmentalists argued that past permitting mistakes were no excuse to ignore the current crisis.[29]

The battles were fought at the political, legal, and personal levels. In 1996, Pinellas County spent $800,000 on a campaign to convince the public that a drought rather than groundwater pumping was responsible for the environmental harm to lakes and wetlands.[30] At one point, Pinellas County even sued citizen activists whose wells had gone dry in Pasco County, in an attempt to intimidate and silence them.[31] Over the years, litigation cost taxpayers more than $10 million, "with not one new drop of water served to the public," observed Honey Rand, author of a book about the water wars.[32]

State senators threatened that if local leaders couldn't resolve the issues, the lawmakers would step in and do it for them—something no one in the region wanted. In 1998, after more than two decades of battle, a truce was finally declared. The three counties and the cities of St. Petersburg, Tampa, and New Port Richey signed a six-party "interlocal agreement" to work together. The regional water authority was reorganized as Tampa Bay Water, a nonprofit special district of the State of Florida to provide wholesale water to the municipalities that provide drinking water in the Tampa Bay region. Tampa Bay Water became the largest water utility in Florida and one of the largest in the southeastern United States. The utility was required to reduce groundwater pumping from eleven wellfields from 192 to 90 million gallons a day within a decade.[33] Water conservation and new water supplies would make up the difference.

The new water supplies were met by a large reservoir to store water harvested from local rivers at high flow, pumping groundwater from relatively nonimpacted areas, and construction of a desalination plant. Plagued by bankruptcies and technical problems, the desalination plant became fully operational in late 2007, years behind schedule and $40 million over budget. It was the largest in the Western Hemisphere until eclipsed by San Diego's desalination plant in 2015.

Potable reuse was not part of Tampa Bay Water's plans for new drinking-water supplies. The wastewater-treatment plants were managed by others, and Tampa Bay Water clearly favored water sources under its control and influence. The push for potable reuse would come from counties and cities. The Southwest Florida Water Management District, which has Florida's largest reuse program, would provide financial support.[34] Meanwhile, some of the wounds and mistrust from the multi-decadal water wars remained beneath the surface.

Like Denver and San Diego, Tampa was one of the early experimenters with potable reuse. The city operated a pilot facility in the 1980s as part of the Tampa Water Resource Recovery Project or TWRRP. (Some called it "Twerp.") By 1998, when the interlocal agreement was signed, the city appeared to be less than a year from obtaining permits for potable reuse, but the newly formed Tampa Bay Water utility abandoned the project in favor of the desalination plant and new surface-water reservoir.[35]

Tampa's early experience with TWRRP set the stage for its continuing interest in potable reuse. Another motivating factor was the 1972 Wilson-Grizzle Act, which required advanced wastewater treatment for any discharges to Tampa Bay. Tampa and St. Petersburg took different approaches to meet these requirements. A comprehensive system similar to St. Petersburg's pioneering nonpotable-reuse program would have been cost prohibitive for Tampa. While St. Petersburg had wastewater-treatment plants at the four corners of the city, Tampa had a single very large plant serving the entire area that would require much larger pipes and large-scale excavations to distribute recycled water to the area. So the city decided to upgrade its sewage-treatment plant to meet the act's discharge requirements.[36]

The Howard F. Curren Advanced Wastewater Treatment Plant (Curren Plant) opened in 1979 as a state-of-the-art facility that sharply reduced nitrogen discharges to the bay through tertiary treatment. Curren, a retired navy captain who was assistant director of the city's department of sanitary sewers, had worked tirelessly to make the plant a reality.[37] The upgrades cost more than $90 million, making it more expensive than Tampa International Airport at the time. When it opened, officials celebrated by sipping treated effluent from champagne glasses.[38] More importantly, Tampa Bay made a remarkable recovery.[39]

Meanwhile, Tampa dipped its toes into water reuse. By the early 2000s, the STAR (South Tampa Area Reuse) project brought reclaimed water to selected neighborhoods, but the city council rejected plans to expand the project as too expensive.[40] Residents also were unhappy about the prospect of more chewed-up streets and yards.[41] In 2009, Tampa International Airport started using reclaimed water for landscape irrigation and cooling towers. However, most of the treated wastewater, in excess of fifty million gallons a day, continues to be discharged into Tampa Bay. The amount of this lost water is twice the capacity of Tampa's desal plant.

For the past couple decades, Tampa has been looking for ways to make better use of its treated wastewater beyond purple pipes and dumping the rest into the bay. The election of Mayor Bob Buckhorn in 2011 brought new life to the city's quest. Buckhorn's administration evaluated two potable-reuse scenarios through an effort known as the Tampa Augmentation Project, or TAP.

The utility first explored the idea of piping treated effluent from the Curren Plant to a site where it would recharge rapid infiltration basins and/or wetlands. From there the water would make its way to the regional surface-water supply. Both recharge options were rejected. The site for rapid infiltration basins turned out to have a thick layer of clayey soils, while the wetlands were rejected because of an otherwise positive feature—they were healthy and did not need the water for rehydration.[42]

A second option explored injecting fifty million gallons a day from the treatment plant about eight hundred feet into the Floridan aquifer. The city would pump an equivalent amount of water back up from a depth of about three hundred feet. Some of the water recovered from the aquifer would be sent directly to Tampa's drinking-water treatment facility. The rest would go into the Hillsborough River Reservoir, the city's primary source of drinking water, where it would enhance the regional water supply and help meet downstream minimum flow requirements in the river. Tampa is unique among major municipalities in the Tampa Bay region in having its own reservoir for surface-water supply since the mid-1920s.

Buckhorn was enthusiastic about the prospects, saying, "I think it's a project that not only would guarantee Tampa's drinking supply but would be hugely helpful for the environment. It's one of the things I'd really like to get done—or at least started—before I leave." He viewed it as "a nice legacy."[43]

The reuse project faced headwinds from several groups. The League of Women Voters and environmental activists opposed it, concerned about costs, contaminants of emerging concern, and environmental impacts. Some called it toilet-to-tap rather than TAP. Tampa Bay Water and St. Petersburg viewed Tampa's attempt to become self-sufficient in its potable water supply as a threat to regional cooperation. Tampa Bay Water officials argued that the 1998 interlocal agreement had made them the sole provider of new drinking-water sources for the three-county area. St. Petersburg representatives were concerned that Tampa might use the project to gain independence from Tampa Bay Water, leaving water-scarce St. Petersburg in the lurch. A Hillsborough County commissioner, who sat on Tampa Bay Water's board, described an executive committee meeting on the topic in 2017 as "three hours of hell."[44]

Buckhorn never achieved his goal.

TAP has been replaced by a new proposal—PURE for "Purify Usable Resources for the Environment."[45] As the name suggests, more emphasis is now placed on water for the environment to preserve ecological health and diversity. Similar to the second TAP alternative, purified wastewater will be pumped into the Floridan aquifer via a series of recharge wells. Freshwater recovered from a set of shallower wells will then be discharged to the Hillsborough River Reservoir.

The project would have three key benefits: It will create a saltwater intrusion barrier to help safeguard Sulphur Springs and other freshwater resources that have become increasingly saline. It will provide freshwater to keep reservoir levels high and meet minimum flow requirements for the Lower Hillsborough River. And it will position Tampa to address legislation that requires cities to eliminate nonbeneficial surface-water discharges of treated effluent by January 2032.[46]

Until recently, the city has relied on water from Tampa Bay Water only during drought conditions when it's short on river water. With increasing demands, Tampa is now purchasing water during normal as well as drought conditions.[47] Not only would the reuse project eliminate the city's needs for wholesale water from the Tampa Bay Water system, but there could also be water to spare for Tampa Bay Water to offset their groundwater pumping. The reuse project might also replace other water sources that are currently used to comply with the Lower Hillsborough River minimum flow requirements.

Overall, the city of Tampa is sitting on a potable-reuse gold mine. It will take a good deal of public education and collaboration before any plan can be put into place. Recognizing this need, the Tampa Water Department is meeting with environmental and community groups on a regular basis and assembled a third-party advisory panel to provide technical advice.[48] Resolving the issues with Tampa Bay Water remains a work in progress.

Hillsborough County, a nationwide leader in nonpotable reuse, is also actively pursuing potable reuse. The Hillsborough County Public Utilities Department has the largest retail reclaimed-water program in the nation, says Bart Weiss, chief officer of innovation and resiliency for the county.[49] While reclaimed water is often fully utilized during the dry season, much of it is discharged into Tampa Bay during rainy months, when water for irrigation is not needed. The Hillsborough utility ultimately plans to make 100 percent beneficial use of its reclaimed water. To achieve this ambitious goal, additional uses are needed during wet periods.

The utility is targeting a fast-growing area east of Tampa Bay, where groundwater pumping has caused large-scale saltwater intrusion and severely impacted lake levels and spring and river flows. This area has the dubious distinction of being designated the "Most Impacted Area" within the Southwest Florida Water Management District.

Conceptually, the South Hillsborough Aquifer Recharge Project (SHARP) is a straightforward idea. During the wet season or other times of excess, reclaimed water from the Hillsborough utility is injected into a coastal zone of the Floridan aquifer that is already contaminated by saltwater intrusion. The recharge through this linear well system creates a mound or barrier that prevents saltwater from intruding further into the freshwater portion of the aquifer. This mounding effect is greatest at the wells but also increases aquifer levels for several miles inland, providing for future groundwater pumping by Tampa Bay Water. To achieve a net benefit for the regional aquifer, withdrawals will not exceed 90 percent of the recovery realized. For example, if recharge is ten million gallons per day, then Tampa Bay Water purchases nine million gallons per day of withdrawal credits, and one million gallons per day remains in the aquifer. The more Hillsborough recharges, the more the region benefits. Tampa Bay Water is not the only winner. With increasing water levels, other existing aquifer users in the area and the environment benefit from rising groundwater levels. The Tampa Bay estuary also benefits from reductions in nutrient loads. Long-term goals are for ten million gallons a day by 2028, potentially expanded to twenty million gallons by 2040.

Hillsborough County is pursuing another novel use of reclaimed water—creation of an estuarian habitat for juvenile fish on the shores of Tampa Bay. An abandoned tropical fish farm containing several hundred breeding ponds is being transformed into a pond receiving reclaimed water and stream terminating into a tidal pool with mangrove islands. Since abandonment, the shallow ponds where the tropical fish were grown have been overtaken by invasive species. Using reclaimed water, the site will be restored to a diversity of habitats from uplands to very low salinity areas to intertidal lagoons and islands. The goal is to meet the specific needs of recreationally important fish, including tarpon, snook redfish, and sea trout. "It's the first time I know of that a water utility and environmental scientists work together using reclaimed water to help fish and our bay ecosystems," says Weiss. "There are other places that discharge into grassy wetlands that may have a similar impact, but they haven't focused on ecosystems as much as we have."[50]

Hillsborough County is also exploring direct potable reuse. In 2016, a small pilot project (Florida's first) produced about a thousand gallons of purified water over the course of a week. After testing for contaminants, the county followed the script of what appears to be a sure-fire way to generate interest and good publicity (as we'll see in chapter 11)—reaching out to craft beer brewers. Special Hoperations, a beer-brewing club located in Tampa, identified about one hundred registered home brewers. Each brewer received two five-gallon buckets of purified water to make their brew. The winners were selected at a People's Choice tasting contest during the 2016 annual meeting of the WateReuse Association, which was held in Tampa that year.[51]

Several other Florida communities are considering potable reuse to meet future needs. Tampa Bay's third largest city, Clearwater, has explored the idea of injecting advanced treated wastewater into an untapped zone of the lower Floridan aquifer. Water would be withdrawn from a second well in the upper aquifer zone, treated, and delivered to residents' homes. It would take about ten years for the water to migrate from the lower zone to the upper zone. The goal was to reuse about a third of the city's wastewater in this way. In 2019, after ten years of study and a $6.2 million investment, Clearwater completed the final design and permits to break ground, but higher-than-expected construction and operation costs have delayed the project indefinitely.[52]

The suburban city of Altamonte Springs, near Orlando, also has long been on the vanguard of water reuse in Florida. In the 1980s,

Altamonte Springs retrofitted neighborhoods and developments to deliver reclaimed water to almost every property in the city for irrigating lawns and greenspace.[53] More recently, the city aspires to provide about 5 percent of its future water demand with direct potable reuse. A pilot project, exploring an ozone-biofiltration process that avoids the use of reverse osmosis, served the dual purpose of evaluating the technology and as an educational platform about the benefits of potable reuse. Known as pureALTA, the project was recognized as the Innovation Project of the Year by the WateReuse Association in 2017.[54]

Florida now seems to be approaching its fourth era of water reuse—the "Age of Potable Reuse." There are plenty of incentives. Many of the state's fresh groundwater and surface-water resources are tapped out. Turning to potable rather than nonpotable reuse avoids the need for expensive purple pipes and the disruption caused by their installation in developed areas. The seasonality of demand for irrigation water severely limits the ability of purple pipes to fully use their capacity. Potable reuse also would reduce nutrient loads to Florida's sensitive rivers, lakes, and estuaries.

But challenges remain. As Clearwater demonstrates, cost is a definite consideration. Cities are also battling public perception and environmental concerns. A tug-of-war exists between those who see the need for more water for a growing population and those concerned that providing more water just fuels more growth. Many argue that more attention needs to be directed to conservation for meeting future needs. Of course, conservation and reuse are complementary. The state also has large reserves of brackish groundwater.

Toilet-to-tap concerns almost inevitably resurface in debates over potable reuse. In 2018, the Florida legislature passed a bill that encouraged expanded use of recycled water to replenish the state's aquifers. The bill passed with overwhelming bipartisan support but was opposed by a group of environmentalists who called it "poopy water." Republican Governor Rick Scott vetoed the bill. Scott was planning to run for the Senate and didn't want to alienate environmentalists any further than he already had on other matters. A leading opponent of the bill threatened, "If he lets this bill become law, he knows he's going to get a new nickname: Gov. Poopy Water. Maybe he doesn't want that."[55] Scott's veto was a temporary setback. A similar bill passed in 2021.

Among its provisions, it declares potable reuse as an alternative water supply—a major boost for the eligibility of potable-reuse projects for funding.[56]

In another major development, the Florida Potable Reuse Commission was created to develop a consensus-based framework to advance potable reuse in Florida. An extensive two-year process brought together associations that represent water and wastewater utilities and stakeholders representing agriculture, the environment, public health, and associated industries. All meetings were open to the public. The commission released its report in 2020 with positive recommendations to advance potable reuse in Florida, including indirect potable reuse using groundwater and direct potable reuse.[57] The challenge now is to act on the recommendations—the Age of Potable Reuse depends on it.

Chapter Nine

Microbes and Natural Buffers

Always drink upstream from the herd.

—Will Rogers

At the opposite end of the treatment continuum from highly purified water is so-called "raw water" that receives no treatment *on purpose*. Raw-water proponents swear by its health benefits. We're not talking about private well water, but rather untreated spring water packaged as expensive boutique water. "It has a vaguely mild sweetness, a nice smooth mouth feel, nothing that overwhelms the flavor profile," claims one store manager.[1] Critics are flabbergasted that anyone would pay $36.99 for glass orbs containing 2.5 gallons of water advertised as unfiltered, untreated, and unsterilized.

"In an age where 'unprocessed' or 'raw' foods are considered to be healthier; some people have extrapolated that concept to drinking water," observes Seth Kellogg, a consultant with the National Ground Water Association. She cautions that there's a reason most water is treated—the sources are not clean enough for humans to drink it safely.[2]

Adherents to raw water share deep distrust of tap water, particularly the fluoride added to it and the lead pipes some of it passes through. They also contend that the wrong kind of filtration (of both tap water and bottled water) removes beneficial minerals. They say advanced water-treatment techniques, such as reverse osmosis and ultraviolet light, kill healthful bacteria known in raw-water parlance as "probiotics." To emphasize the presence of these probiotics (and because "raw water" sounds too much like "raw sewage"), many proponents prefer the term *live water*.

"Sadly, the probiotics in water won't necessarily help fend off any diseases," notes Dr. Morton Tavel, author of the book *Health Tips, Myths and Tricks: A Physician's Advice*. He adds that although some probiotics may be beneficial to health, you can obtain these organisms in a far safer fashion from products such as cultured yogurt. Tavel calls the raw water fad "one of the most ridiculous ideas I have ever heard."[3] Like the movement against vaccines, the movement has brought together unlikely allies from the far left and the far right. Conspiracy theorists like Alex Jones, founder of the right-wing website *Infowars*, have long argued that fluoride is added to water to make people more docile.[4] Raw water aficionados appear to have little understanding of the ease with which waterborne disease is transmitted and how lucky we are compared to the more than two billion people worldwide who drink feces-contaminated water without all the protections afforded to them by treatment.[5] Fortunately, microbial contaminants (pathogens) are taken very seriously by the wastewater and drinking-water community.

Both chemicals and pathogens threaten drinking water supplies, but the threats they pose occur at different time scales. For the vast majority of chemicals, long-term (chronic) exposure levels are the principal concern. Short-term fluctuations in concentrations are much less important.[6] In contrast, a single exposure to pathogens can cause serious illness. As such, pathogens deserve special attention anytime treated wastewater comes in contact with humans.

The vast majority of microorganisms found in water do not cause disease. The primary culprits are enteric pathogens that replicate in the intestinal tract of humans or animals and are spread by fecal-contaminated water. Any potable water supply receiving human or animal wastes can be contaminated by pathogens. This fecal-to-oral connection is the primary reason why fear of recycled wastewater is so prevalent.

Pathogens in drinking water come in many shapes and sizes. Bacteria, viruses, and parasitic protozoa differ in their occurrence, health effects, and resistance to water treatment. Let's take a quick look at each.

Bacteria are ubiquitous. These microscopic organisms can cause disease, but they also play an essential role in breaking down human, animal, and plant wastes so that life on this planet can continue. Thanks to disinfection of drinking water, many of the worst of the lot—those bacteria that cause deadly waterborne diseases such as cholera, typhoid, and dysentery—have been virtually eliminated in the United States. Other pathogenic bacteria, such as some E. coli, Campylobacter, and Salmonella, continue to cause waterborne disease outbreaks. Symptoms include diarrhea, fever, coughing, and vomiting as our immune system tries to rid the body of these infectious organisms.

Viruses are of particular concern in water because of their small size (typically 0.025 to 0.3 microns) and resistance to disinfection. Their simple structure—a protein coat surrounding a core of genetic material (DNA or RNA)—allows prolonged survival in the environment. These tiny pathogens can be excreted in enormous numbers (trillions) in small amounts of feces from infected people. In some cases, only one to ten viruses of the many trillions excreted is needed to cause acute gastrointestinal illness—vomiting, diarrhea, and potentially death. This relation between high numbers excreted and very few needed to cause illness is a primary reason why virus-related illness is so easily transmitted. Chlorine is less effective at killing viruses than bacteria.

Protozoan parasites, such as *Giardia* and *Cryptosporidium*, are frequently transmitted through water in environmentally resistant forms, known as cysts and oocysts. *Cryptosporidium* can cause severe watery diarrhea and can be fatal for vulnerable populations (infants, the elderly, and people with compromised immune systems, such as those who have AIDS). In 1993, *Cryptosporidium* gained notoriety as the cause of the largest reported drinking-water outbreak in U.S. history. More than four hundred thousand people in Milwaukee, Wisconsin, were infected, and over fifty people died. The exact source and cause of the outbreak are still unknown. However, the general consensus is that rivers swelled by spring flooding caused human sewage and other wastes to transport *Cryptosporidium* oocysts into Lake Michigan, and from there to the intake of one of Milwaukee's drinking water plants.[7] Almost three decades later, the incident remains the poster child for the importance of rapid detection and response to outbreaks that threaten water systems.

Cryptosporidium were a relatively unknown threat to drinking water at the time, making them a lesson in the possibilities of unknown pathogens with unique properties entering drinking-water systems—with or without water reuse.

Cryptosporidium is resistant to conventional drinking-water chlorination and must be specifically filtered or inactivated by ultraviolet radiation, ozonation, or other means. After the Milwaukee outbreak, the U.S. Environmental Protection Agency (EPA) developed regulations to identify high-risk drinking-water systems and take actions to reduce the risks associated with *Cryptosporidium*. These actions also reduce the risks to *Giardia*, which often co-occurs.[8]

Pathogens do not persist indefinitely in the environment. Researchers commonly measure the rate at which their populations decline in half-lives—the time it takes for a 50 percent reduction in number. Half-lives of pathogens vary from a few hours to many months, depending on the pathogen and environmental conditions such as temperature and pH.

With few exceptions, potable-reuse projects worldwide include an environmental buffer (an aquifer, reservoir, or constructed wetlands) between the wastewater discharges and the intake to the drinking-water treatment plant. These environmental buffers provide the benefits of dilution and attenuation of contaminants by biological, chemical, and physical processes—as well as time to take corrective action in the event of treatment plant failures and surprises. They also potentially serve a psychological function by reducing people's mental association of the drinking water with sewage. This last benefit may be overrated, given the controversies that persist with some indirect potable-reuse projects.

Most indirect potable-reuse projects use groundwater as the environmental buffer. Among its advantages, groundwater can be an excellent way to store water until it is needed. "Groundwater banking" has some distinct advantages over traditional surface reservoirs. Many of the best sites for dams are already taken, construction costs are high, and considerable controversy often surrounds the environmental effects of tampering with the natural flow of rivers. One of the most compelling arguments is that storing water underground avoids massive losses to evaporation, particularly in hot, dry climates.

The Eastern Municipal Water District (EMWD) in Riverside County, California, is a good example of the integrated use of groundwater (fresh and brackish) and recycled water. EMWD supplies water to more than 850,000 people in one of the fastest-growing areas in California. Half of the water supplied in this semi-arid inland district is imported. The other half comes from three sources: recycled wastewater (35 percent), fresh groundwater (10 percent), and brackish groundwater (5 percent).[9]

EMWD began marketing reclaimed water to local farmers for irrigating feed and fodder crops in 1966. As the population grew, the reuse program expanded to include public landscaping, industrial facilities, and environmental enhancement of wetland areas. Today, EMWD is one of the largest-by-volume water recyclers in the country, as well as one of the few agencies that consistently achieves 100 percent beneficial use of its recycled water. Treatment facilities and storage ponds ensure year-round water availability of the recycled water. Recently, EMWD embarked on an ambitious initiative, known as Groundwater Reliability Plus, to bank imported water underground for drought proofing and future growth. The agency is also exploring indirect potable reuse using groundwater as the environmental buffer, as well as expanding its use of desalters (using reverse osmosis) to develop more of its brackish groundwater.[10]

Underground storage comes with its own set of challenges. First and foremost, a suitable aquifer is required to store the water, which means that you need the right geology. The water must also remain of a suitable quality. For example, in some geologic settings, the chemical interactions of recharged water with the rock matrix can release arsenic or other toxic elements. In addition, spreading basins and injection wells often clog during their operational life and require specialized knowledge to maintain long-term operations.

California's regulations for well injection place the burden of pathogen reduction on advanced wastewater treatment, including reverse osmosis and advanced oxidation processes. This is called "full advanced treatment" or *fat* to the chagrin of some water officials. Conversely, spreading basins rely heavily on soil-aquifer treatment for pathogen reduction.

Soil-aquifer treatment reduces pathogens through several mechanisms. Pathogens can be consumed by other organisms in the subsurface. They can become attached (sorbed) to particles, which removes them from the water—or at least delays their transport. And finally, they can be filtered out when they're too large to fit through the aquifer pores and cracks. The extent of filtration depends on the type of soil and rocks through which groundwater flows. For example, silts are more effective at trapping microorganisms than sands, and sands are more effective than gravel. The extent of filtration also depends on the size of the organisms. Protozoa, such as *Giardia* or *Cryptosporidium*, are larger and are typically trapped in common aquifer materials. Bacteria fall somewhere in between viruses and protozoa in size and can travel through groundwater under certain conditions. Viruses readily pass through many pores and cracks but are more susceptible to adsorption as influenced by sediment particle size, organic carbon content, pH, and other factors.

California has the most comprehensive water-reuse regulations of any state. The first regulations for groundwater replenishment with recycled water in 1978 focused on organic chemicals and, for the most part, considered water-recycling projects on a case-by-case basis.[11] After years of development, comprehensive regulations for groundwater recharge and surface-water augmentation were issued in 2014 and 2018, respectively. As with many regulations, these are spelled out in mind-numbing detail full of regulatory jargon. Let's dip our toes into the regulatory pool to get a sense of what's involved for pathogens.

The basic idea is to apply credits for pathogen removal by different treatments, a concept that originated with the 1989 EPA Surface Water Treatment Rule. The original EPA rule and subsequent updates lay out pathogen removal/inactivation requirements for enteric viruses, *Giardia lamblia*, and *Cryptosporidium*.

The ability of a treatment process (including soil-aquifer treatment) to remove pathogens is measured by taking the logarithm of the ratio of the pathogen concentration before and after the treatment process or treatment train. This is referred to as the log reduction value (LRV). While this may sound cryptic, you don't have to pull out a math book or computer to see how it works. For integer log reductions, just count the number of 9's. A 1-log reduction means a 90 percent reduction in the number of the target pathogen(s). A 2-log reduction means a 99 percent reduction. A 3-log reduction means 99.9 percent reduction, and so forth. Removal rates from each step in a treatment train can simply be added up using this system: a 1-log reduction by one process and 3-log reduction by another yields a 4-log reduction.

A key part of California's regulations for indirect potable reuse is the *12/10/10* rule for pathogens. The rule requires 12-log reductions of enteric viruses (that's a 99.9999999999 percent reduction!) and 10-log reductions of both *Cryptosporidium* oocysts and *Giardia* cysts. These are the estimated reductions needed to achieve a risk of one infection per ten thousand people per year, a common risk level for pathogens.

The way it works is that a facility gets LRV credits for different natural and engineered barriers. Water recharged through spreading basins must first undergo at least tertiary treatment with disinfection. Soil-aquifer treatment is relied on to complete the process of pathogen inactivation and removal.

Different credits apply for protozoa and for viruses. Any project that provides a minimum six-month retention time after spreading gets a 10-log reduction credit for both *Cryptosporidium* and *Giardia*, thereby receiving full credit for removing these parasitic protozoa under the regulations. Viruses receive fewer LRV credits, because they travel much easier through groundwater than protozoa. A 1-log credit for enteric virus reduction is received per month of aquifer retention time. To meet the standard of 12-log reduction of viruses, the log reductions must come from at least three treatment steps.

All of this depends on determining the aquifer retention time to the regulatory agency's satisfaction. Early draft regulations simply assumed that a setback distance of five hundred feet from the edge of the spreading grounds to the nearest drinking water well would provide at least six months of retention time. However, it was soon apparent that a simple distance criterion was not sufficient—well depth and aquifer characteristics are often more important.[12] The retention time must be demonstrated using tracers or modeling. Full credit only applies if an added tracer is used to demonstrate the retention time; lesser credit is given for modeling studies.

California is currently working on completing regulations for direct potable reuse by 2023. An expert review panel was formed early on to provide advice in developing criteria. As part of initial steps, the expert panel identified three research priorities for pathogens.[13]

The first research priority is to use quantitative microbial risk assessment (QMRA) to confirm the LRVs required to achieve a risk of one infection per ten thousand people per year. QMRA estimates the potential risks to human health by taking into account the range of

concentrations of infective pathogens in the source water, the effectiveness of treatment barriers in removing or inactivating these pathogens,
and the risk of infection from drinking the treated water. QMRA can
evaluate the impact of different types of failure on risk. For example,
what if the ultraviolet (UV) system goes down for some period of time?

Conservative assumptions are typically used in applying QMRA. For
example, adenovirus is known to be resistant to UV disinfection and so
is a conservative indicator of UV effectiveness. If UV doses are effective against adenovirus, then they should be more effective on more
vulnerable viruses. As another example, California utilized the highly
infectious rotavirus in establishing regulatory criteria for indirect potable reuse. If the reduction of rotavirus provides a sufficient safety factor
in the probability of an infection for a certain exposure, then similar
reductions of other less infectious viruses also should be satisfactory.[14]

The second research priority identified by the expert panel is to
obtain more complete information on pathogen concentrations and their
variability in raw wastewater—a critical factor in addressing the first
research priority. The third research priority is the most challenging—to
collect data on pathogen concentrations in raw wastewater associated
with community disease outbreaks. Outbreaks of emerging pathogens are of particular concern, as evidenced by the recent worldwide
COVID-19 pandemic.

Fortunately, the coronavirus that causes COVID-19 was not a threat
to drinking water. Because of their fragile fatty envelope, coronaviruses
are less persistent in the environment and more sensitive to treatment
than the enteric viruses targeted in potable reuse. Conventional water-
and wastewater-treatment methods removes or kills coronaviruses.
Although not a drinking-water threat, COVID-19 illustrates the possibilities for outbreak monitoring using molecular techniques to evaluate
and characterize waterborne pathogens.

The most widely used technique, polymerase chain reaction (PCR),
rapidly replicates a specific DNA fragment in a simple enzyme reaction. Quantitative PCR (qPCR) enables detection and quantification of
the target sequence in real time. The use of qPCR and similar methods makes it possible to characterize a broad spectrum of potential
pathogens.

Early scientific studies during the pandemic demonstrated that PCR
could frequently detect the genetic material of the virus in the feces

of infected individuals. This finding suggested that the genetic signal in wastewater could provide an early, cost-effective community-level indicator of the presence of COVID-19. (Note that PCR cannot distinguish between infectious and inactivated viruses, but other evidence indicates that the presence of infectious virus appears to be small, if not negligible in the feces.[15])

Wastewater surveillance is not a new concept. It has been used for poliovirus surveillance as part of the World Health Organization polio-eradication program, as well as to investigate opioid and other drug use in communities.[16] Researchers in the Netherlands were the first to report that they had detected the COVID-19 virus in wastewater samples.[17] Almost overnight, a global research effort was spawned on wastewater surveillance. By sampling the wastewater, researchers detected COVID-19 hot spots days before those cases appeared in hospital admissions data and clinical testing.[18] Additionally, the method is sensitive to both symptomatic and asymptomatic cases and doesn't require testing of individuals to identify a problem. As one wastewater manager put it, "Sewage sleuthing to provide early warnings about COVID-19 outbreaks is the most exciting thing to happen in my field in a long time."[19] Opportunities for similar early warning systems applied to potable reuse is an active research area.

SCOTTSDALE, ARIZONA

Scottsdale, Arizona, an affluent community and golfing mecca in the Greater Phoenix area, is home to Arizona's first advanced wastewater-treatment plant. This treatment facility relies on an unusual environmental barrier for indirect potable reuse—vadose zone wells (also known as dry wells). The story of this relatively unique approach to an environmental buffer, as well as Scottsdale's leadership in water reuse, begins with groundwater overdraft accompanying rapid population growth after World War II.

By the 1970s, groundwater depletion in south-central Arizona from Phoenix to Tucson was impossible to ignore. In many areas, groundwater levels had declined hundreds of feet, often accompanied by land subsidence and earth fissures. The need to aggressively manage the state's finite groundwater resources was obvious but controversial, particularly among the agricultural community. After years of debate, Arizona passed the landmark Groundwater Management Act in 1980.

The act created five Active Management Areas (AMAs) that cover much of the state's population and groundwater use, including a large area around Phoenix. The act required each AMA to develop plans to wean itself from overreliance on groundwater pumping. The Phoenix AMA's goal is for annual groundwater withdrawals not to exceed the annual rate of aquifer replenishment by 2025—a condition referred to as "safe yield."[20] In addition, new housing developments in AMAs must demonstrate an assured water supply lasting at least one hundred years, highlighting the need for renewable water supplies such as Colorado River water and treated wastewater.

Passage of the Groundwater Management Act was effectively a quid pro quo for receiving federal funding for the Central Arizona Project (CAP) to bring Colorado River water to Phoenix, Tucson, and other parts of south-central Arizona. The bottom line was that the federal government wasn't going to fund construction of the CAP without hard evidence that Arizona would get its groundwater pumping under control.[21]

The Groundwater Management Act and CAP provided strong incentives for water recycling and for storing water underground, actions in which Arizona had already begun to establish itself as a leader. Beginning in the mid-1960s, Herman Bouwer with the U.S. Department of Agriculture in Phoenix and Sol Resnick at the Arizona Water Resources Research Center in Tucson carried out pioneering work on soil-aquifer treatment.[22] In 1972, the state issued some of the first rules governing reclaimed water in the nation. Today, more than 80 percent of all treated wastewater generated within the Phoenix AMA is beneficially used for various purposes or recharged.[23] Among the more notable uses, the Palo Verde Nuclear Generating Station, the country's largest nuclear power plant, uses 100 percent reclaimed water for cooling.

With water available from the CAP, the state began an ambitious program to store its unused Colorado River allocation underground and went about developing a comprehensive regulatory framework for groundwater storage and recovery.[24] Overdraft of the state's aquifers had provided plenty of space for underground storage.

Water conservation also became a high priority in the AMAs. Phoenix uses the same amount of water as it did twenty years ago, despite adding four hundred thousand more residents. "In 2000, some 80 percent of Phoenix had lush green lawns; now only 14 percent does," says Kathryn Sorensen, Director of Phoenix Water Services.[25] Over the past couple decades, recurring droughts and heightened awareness of climate change have increased concerns about reductions in the availability of Colorado River water. As a result, wastewater recycling, water banking in underground reservoirs, and reducing per-capita use have taken on even greater urgency.

Prior to the Groundwater Management Act, Scottsdale was completely dependent on groundwater for its water needs, and most of its wastewater went to Phoenix for treatment. The city made a small venture into water recycling in 1981, using tertiary-treated wastewater to irrigate the Gainey Ranch Golf Course.

In 1998, the city completed the Scottsdale Water Campus to reduce the city's dependence on groundwater and make fuller use of its wastewater. The campus includes a drinking-water treatment plant, a tertiary water-reclamation plant, an advanced wastewater-treatment plant, and a state-of-the-art water-quality laboratory. It also has the look and feel of a campus, being designed to aesthetically blend into the local environment. The advanced treatment facility treats the tertiary effluent through ozonation, ultrafiltration, RO, and UV disinfection. Recycled water from the Water Campus serves two end uses: irrigating golf courses and aquifer recharge.

Through a public-private partnership, Scottsdale Water provides recycled water for turf irrigation to twenty-three golf courses in north Scottsdale. For their part, the golf clubs invested in the advanced-treatment facilities and help pay for expansions along with their share of the operating costs. When irrigation demand is high, the golf courses receive recycled water. Tertiary effluent is blended with advanced-treated water, as necessary, to keep dissolved solids (particularly sodium levels) low. When irrigation demand is lower, Scottsdale Water recharges the groundwater with advanced-treated water.[26]

Getting the treated wastewater to the deep unconfined aquifers is no small challenge. Sky-high real-estate prices and the scarcity of appropriate locations hamper the use of spreading basins. Direct-injection wells would have to be five hundred, or more, feet deep. Scottsdale turned to vadose-zone wells as a cost-effective alternative. (The vadose zone is the material from the land surface to the water table.) In this method, water is injected into a series of dry wells about 180 feet deep. This provides a jump start from which the water flows through several hundred feet of vadose zone before reaching the water table and mixing with local groundwater.

Vadose-zone wells have some distinct advantages. They are relatively inexpensive compared to direct-recharge wells, do not require the extensive space of recharge basins, have minimal evaporative losses, and can be placed in a wide range of locations. A disadvantage is that vadose zone wells can't be pumped for "backwashing" or scraped like spreading basins to control clogging. Scottsdale Water has maintained their vadose-zone wells through careful design and operation as well as the use of highly treated water for recharge.

Disposal of the brine from RO is another major challenge. Fortunately, Scottsdale has long collaborated with Phoenix and three other nearby cities. The five cities collectively own the 91st Avenue Wastewater Treatment Plant, the largest of its kind in the Southwest, where the brine can be sent.[27]

In 2006, Scottsdale became the first city in Arizona to deposit more in its groundwater bank account than they were taking out, thereby achieving "safe yield" almost twenty years before required by the Groundwater Management Act for the Phoenix AMA. The Water Campus played a significant role in this achievement, along with conservation and the availability of surface water from the Central Arizona Project (and to a lesser extent, the Salt River).

In 2019, Scottsdale Water further burnished its water-reuse profile when it became the first facility in Arizona—and only the third in the nation—to be permitted for direct potable reuse (DPR). Scottsdale Water has no immediate plans for DPR, as indirect potable reuse using groundwater as a reservoir is better suited to the city's large seasonality of demand and long-term storage for future use. Scottsdale Water executive director Brian Biesemeyer explains, "We pursued the DPR permit for demonstration purposes and to help other water providers define their paths toward optimal water reuse."[28]

Chapter Ten

Contaminants of Emerging Concern

You shouldn't have cancer-causing substances in the food supply, unless people like them a lot.

—Donald Kennedy, U.S. Food and Drug Administration Commissioner, commenting on failure of Congress to embrace a proposed ban on the artificial sweetener saccharin in the 1970s (more recent evidence suggests it's not a human carcinogen)[1]

In the late 1990s, scientists in Germany reported the common presence of pharmaceuticals in drinking water and wastewater-treatment plants.[2] Then in 2002, the U.S. Geological Survey released the results of a study that tested 139 streams in thirty states for pharmaceuticals, hormones, and other wastewater contaminants. One or more of these contaminants were detected in 80 percent of the samples. Though the amounts were small (mostly less than one part per billion), almost all of the ninety-five contaminants were detected at least once. The most frequently detected compounds included steroids, caffeine, nicotine metabolites, nonprescription pain relievers, and DEET, the active ingredient in many insect repellents. Antibiotics were detected in more than half the samples. The samples were collected just downstream of urban and agricultural wastewater sources because scientists knew that's where the contaminants would most likely occur in the environment. Although the study targeted likely hot spots, the widespread detection of these compounds was cause for concern. It became one of the most cited environmental science studies of the decade.[3]

These and other studies ushered "contaminants of emerging concern" into the common lexicon. An emphasis on the "emergence" of new contaminants, however, dates back to Rachel Carson's 1962 book *Silent Spring*.[4] Notably, concerns about some contaminants have been "emerging" for a long time. A key attribute of contaminants of emerging concern is the lack of federal drinking-water regulations.

Contaminants of emerging concern are a moving target. Millions of chemicals are potentially present in recycled water, and information about them is continually evolving.[5] Increasingly sensitive laboratory detection methods continue to bring new chemicals to the forefront. These detections, often at parts per trillion or even parts per quadrillion concentrations, greatly exceed our understanding of their health and ecological significance. Compounding the challenges, metabolites and transformation products that form in the environment and during treatment may be more toxic than the parent compounds. And perhaps most troublesome, very little is known about how mixtures of chemicals, such as a cocktail of tiny amounts of multiple drugs, might increase their toxicity.

The quote at the beginning of this chapter illustrates some key challenges. As sarcastically noted by Donald Kennedy, there's a tendency to politicize decisions about hazardous chemicals. At the same time, the scientific foundation of chemical toxicity is inevitably uncertain, and corrections are often made in light of new findings. Studies in laboratory rats during the early 1970s linked saccharin with the development of bladder cancer, but later studies determined that the bladder tumors seen in rats were due to a mechanism not relevant to humans. Today, no clear link exists between saccharin and human cancer.[6]

About five hundred years ago, the Swiss physician and chemist Paracelsus nailed the basic principle of toxicology when he said, "The dose makes the poison." In other words, just because a chemical is detected doesn't mean it's a health problem. Take ibuprofen, for example. Almost everyone has ibuprofen in their medicine cabinet, because it's so effective at reducing fever and treating pain or inflammation from headaches, toothaches, back pain, and so forth. It's commonly found in wastewater effluent, but this fact by itself doesn't really say much. For example, a California science advisory panel reported 160 parts per trillion as a conservative value in treatment plant effluent.[7] At this concentration, someone would have to drink over five million 8-ounce glasses of water to get a dose equal to one tablet of ibuprofen (200 milligrams). In other words, context is critical.

Placing chemicals in a health-risk context is much easier said than done, especially when you consider the sheer number of substances of potential concern, and what is involved in conducting toxicological and epidemiological studies that consider different populations (infants, elderly, and healthy adults), duration of exposure, other routes of exposure (breathing, eating, or skin contact), and personal traits and habits.

The National Research Council notes that "the very nature of wastewater suggests that nearly any substance used or excreted by humans has the potential to be present at some concentration in the treated product."[8] Among these are painkillers, antidepressants, and other drugs. Americans filled almost six billion prescriptions in 2018—an astounding 17.6 prescriptions per person.[9] Some are used; many are not. Consumers routinely flush unused or expired pills down the toilet. For those pills we do take, up to 90 percent pass through the human body and end up in the water supply.[10] Personal-care products (antibacterial soaps, fragrances, sunscreens, etc.) also find their way into our water supply.

No links have been established between pharmaceuticals at environmental levels in water and adverse effects on human health. It is therefore debatable whether we need to be concerned over the exceedingly low concentrations of pharmaceutical drugs that make their way into drinking water. The long-term risk from any single pharmaceutical at the levels found in drinking-water supplies appears negligible. However, the toxicological effects from chronic exposure to suites of trace pharmaceuticals and other chemicals is less clear.[11]

Conventional wastewater-treatment processes are largely ineffective at removing many pharmaceuticals and personal-care products. Advanced wastewater treatment and soil-aquifer treatment are much more effective, although low concentrations of some compounds may remain even after advanced treatment.

Endocrine-disrupting compounds (EDCs) have perhaps the greatest potential to play into people's fears. These compounds can interfere with hormone functions in the body that regulate such basic features as metabolism, fertility, and brain development, and even our mood and how well we sleep. EDCs come in many forms, including flame retardants (PDBEs), pesticides (DDT), plasticizers (bisphenol A), and synthetic hormones (oral contraceptives). EDCs that can mimic or block the effects of estrogens—the primary female sex hormone—have received the most attention.

Natural and synthetic estrogens have been reported in wastewaters since the 1960s, but it wasn't until wastewater effluents were linked to feminization of male fish that these hormones caught people's attention. The link between concentrations of estrogen hormones in surface waters and feminization of fish is now well established, yet the concentrations in drinking water do not appear to affect humans. For example, the amount of estrogen a person ingests through consumption of dairy milk far exceeds any dose that might be attained through drinking water.

While pharmaceuticals, personal-care products, and EDCs are all of concern, a National Academy of Sciences committee on water reuse concluded that industrial compounds (e.g., 1,4-dioxane) and disinfection byproducts (e.g., NDMA) represent a more serious human health risk than do pharmaceuticals and personal-care products.[12]

Contaminants of emerging concern (CECs) are not restricted to chemicals. More recently, microplastics have emerged as contaminants of concern, although the extent of any dangers to drinking water are presently unknown.[13] Another concern is treated wastewater as a source of antibiotic-resistant bacteria (ARBs) and antibiotic-resistance genes (ARGs). Antibiotic resistance is a serious worldwide public health issue. The broad concern is that pathogens have been developing defenses to antibiotics faster than pharmaceutical companies can develop alternative drugs. Antibiotics, ARBs, and ARGs in wastewater effluents could serve as a contributing factor to growing rates of antibiotic resistance in treating human infections.[14] A combination of secondary wastewater treatment and advanced water-treatment processes is likely to reduce ARB and ARG concentrations in recycled water to levels well below those found in conventional treated drinking water.[15] Nonetheless, the importance of better understanding of ARBs and ARGs in treated wastewater is widely recognized.

The bottom line is that when it comes to contaminants of emerging concern, recycling wastewater has both pluses and minuses. On the downside, many CECs find their way into wastewater. On the plus side, advanced wastewater treatment can remove many of these contaminants to levels below those detected in drinking water.

California has the most rigorous state regulatory program for CECs. In 2010, a science advisory panel developed a list of CECs that should be monitored for potable reuse using groundwater recharge. The idea was to evaluate the effectiveness of advanced wastewater treatment for a representative set of pharmaceuticals, personal-care products, food additives, and hormones. The CECs to be monitored were determined by comparing concentrations of CECs in wastewater with screening levels based on toxicological information, as well as the panel's professional judgment.

The advisory panel selected four health-relevant CECs: 17β-estradiol (a steroid hormone used to treat menopause symptoms and to prevent bone loss in menopausal women), triclosan (an antimicrobial agent found in toothpastes and hand soaps), caffeine (a ubiquitous indicator of human influences), and NDMA (a disinfection byproduct). Periodic reevaluation is planned to add or remove CECs from the list as new occurrence and toxicity information becomes available. In 2018, a reconvened panel recommended that NDMA should be retained while the other three compounds should be replaced by 1,4-dioxane and NMOR (a compound similar to NDMA).[16] A suspected human carcinogen, 1,4-dioxane was commonly used to stabilize chlorinated solvents and is an unwanted byproduct of many consumer products (detergents, cosmetics, shampoos, etc.). It went undetected for many years and is one of the most challenging contaminants to treat.

A margin of safety was built into the panel's risk-based screening framework. According to the panel, the very small percentage of CECs that were recommended for health-based monitoring (3 of 489) reinforced the inherent low potential risk to human health of CECs in recycled water. To make headway in screening for a broader universe of chemicals, the panel emphasized the potential role of two specific *in vitro* bioassays that measure endocrine active chemicals and are sufficiently developed to complement conventional analytical methods. California is adding these two bioassays to its regulations.[17] The search for CECs continues to evolve.

A proactive way to protect against contaminants is to exert controls on what's coming into the wastewater-treatment plant in the first place.[18] The idea is analogous to watershed protection programs for drinking-water plants and water-supply wells. In the case of wastewater, the source areas in need of protection are the network of sewers and sewage collection infrastructure known as the sewershed. Watershed and sewershed protection follow Benjamin Franklin's commonsense advice: "An ounce of prevention is worth a pound of cure."

Prior to the 1972 Clean Water Act, many industries conveniently dumped all kinds of chemicals into the sewer system. In one of the most serious examples, Montrose Chemical Company, the nation's largest manufacturer of DDT, dumped more than 100 million tons of DDT and eleven tons of PCBs into Los Angeles County sewers from 1947 until 1971, when the practice was halted. With minimal treatment, these wastes discharged into Santa Monica Bay. The DDT and PCBs subsequently moved up the food web to fish, birds, and marine mammals and devastated the region's bald eagle and brown pelican populations.[19]

Today, the release of chemicals to municipal wastewater-collection systems is regulated under the Clean Water Act through the EPA National Pretreatment Program, which requires industrial users to obtain a permit restricting what they can discharge to the sewer system. Nonetheless, periodic releases of chemicals can lead to short periods in which elevated concentrations of toxic chemicals enter wastewater-treatment plants. One limitation is that the list of 126 priority pollutants regulated by the National Pretreatment Program has not been updated since its development more than four decades ago. In addition, impurities or byproducts of chemical use tend not to appear in the records of products used by commercial and industrial facilities. And then there are those who operate illegally. For example, in 2020, two brothers were charged with illegally dumping hundreds of thousands of gallons of polluted wastewater from a manufacturer of biodiesel fuel directly into the Stockton, California, sewer system.[20]

Potable reuse increases the need for source control with stringent sewer ordinances and ongoing surveillance, particularly for toxic chemicals that have the potential to pass through advanced-treatment systems. A key goal of source control for potable reuse is to provide a consistent quality of wastewater, which in turn improves operation of the treatment processes.[21] Sewersheds with high contributions of wastewater from industrial, commercial, and medical sources require special attention. Some of these sewersheds simply may not be appropriate for water reuse. These possibilities add support for regulations targeting the use of chemicals that are difficult to remove from wastewater, along with development of biodegradable, nontoxic chemical alternatives (i.e., green chemistry).

Orange County, California, is a good example of proactive sewershed protection. We previously described how the Orange County system tracked down industrial sources of NDMA and 1,4-dioxane. During fiscal year 2019–2020, the Orange County Sanitation District performed 1,422 industrial inspections and collected 3,831 samples to assure the effectiveness of its source-control program.[22] The Orange County Water District also tests for over five hundred compounds in the water it produces, many more than the 126 priority pollutants.

Public education is critical to these efforts. The days when people were encouraged to flush unused medications down the toilet to prevent harm to children have long since passed—but not everyone knows this. Many utilities now have educational and take-back programs for pharmaceuticals and other chemicals. Educational materials stress that other than soap and water, only the 3 Ps—pee, poop, and (toilet) paper—should be flushed down the drain.

Today's most notorious CECs are a large group of chemicals known as PFAS (short for per- and polyfluoroalkyl substances). Because of their resistance to heat, water, and oil, PFAS have been used in a wide variety of commercial and industrial products. Examples include nonstick cookware, water-repellant clothing, stain-resistant carpets and furniture, greaseproof fast-food wrappers, microwave popcorn bags, ski wax, and fire-retarding foam.

Over the past two decades, PFAS have gone from virtually no recognition outside the chemical world to a household name. Among other adverse health effects, exposure to PFAS has been linked to cancer, immune system issues, liver and thyroid problems, harm to developing fetuses or infants, and reduced effectiveness of children's vaccines. PFOA (perfluorooctanoic acid) and PFOS (perfluorooctane sulfonate) have been the most extensively produced and studied, but an alphabet soup of PFAS is used worldwide with thousands of different compounds. Because of their staying power in the environment, PFAS are known as "forever chemicals."

Developing standards for safe levels of PFAS in drinking water has been painstakingly slow. In 2016, the EPA issued a health advisory of 70 parts per trillion for PFOA and PFOS combined. This is a nonenforceable guideline. Some researchers have even suggested that the safe level may be less than 1 part per trillion (ppt). Various analogies are used to imagine such a tiny concentration—a grain of sugar in an Olympic-sized pool, one second in 320 centuries, or a flea on 360 million elephants.

In February 2021, the EPA made a regulatory determination to set drinking-water standards (MCLs) for PFOA and PFOS and is expected to finally issue these within a couple years afterward. The delays in developing federal drinking-water standards for PFAS have caused many states to adopt their own guidelines and standards. California is no exception. In February 2020, the state set response levels for PFOA (10 ppt) and PFOS (40 ppt), with significant repercussions. Forty two of the 195 wells in the Orange County groundwater basin exceeded these levels and were taken offline, requiring more expensive imported water as a replacement.[23]

Reverse osmosis and certain other advanced treatment techniques are effective at removing PFAS, but conventional wastewater-treatment plants are largely ineffective. PFAS "precursors" may even degrade to toxic PFAS during treatment. The PFAS in Orange County groundwater did not come from OCWD's recycled water, but from tertiary-treated wastewater discharges to the Santa Ana River by upstream communities.

Despite having no role in releasing PFAS into the environment, the Orange County water agencies must find ways to remove it from drinking water. In December 2019, the OCWD announced that it was undertaking the nation's largest PFAS pilot program to test treatment technologies for PFAS removal. The district will then construct treatment facilities in each of its member water agencies affected by the chemicals in their well water.[24]

Finally, a note on the value of OCWD's cautionary approach to recycled water. The state does not require the water district to treat all of the water that it recharges at the spreading basins with RO, but the district chose to do so—thereby avoiding its own potential contribution to the PFAS contamination problem.

Chapter Eleven

Achieving Acceptable Risk

Nothing will ever be attempted if all possible objections must be first overcome.

—Samuel Johnson

In May 1998, the city council of El Cajon, one of nineteen sister cities in the San Diego metropolitan area, addressed the topic of recycling wastewater into the drinking-water supply. A retired plumber stood before the city council and spoke into the microphone. "I've spent most of my life trying to separate sewage from water and this is just plain against common sense. In a large, electrical blackout, the electrome-chanical system could shut down and let raw, unfiltered sewage into the drinking water supply." The meeting ended by the council asking for more information about the city's recycling plans.[1]

The following day, in a letter to the editor of the local newspaper, a woman explained that her husband had attended the city council meeting because they were addressing traffic concerns. The issue of recy-cling sewer water had come up, and the council said they needed more information. She wrote, "Why do we need more information? They are suggesting that we bathe and drink water that was previously flushed down the toilets of strangers. What more do we need to know? . . . What makes you think that we know everything there is to know about clean-ing sewer water and making it fit for human consumption? I know that most council members are parents and/or grandparents. Imagine bathing a newborn baby in water that has been recycled from wastewater? The idea is unthinkable."[2]

When asked why she opposed potable reuse, one of the Revolting Grandmas explained, "All I have to do is look in my toilet."[3] This view of the nature of wastewater is commonplace. In reality, wastewater is a mixture of water that has been used for many purposes by homes and companies. Only a small portion of the water that enters a wastewater treatment plant comes from flushing toilets. While we're naturally repulsed by human feces, they're essentially just water plus the food that the body is unable to digest mixed with some of the bacteria that live in the digestive system. These tiny organisms, which are essential to human health and survival, account for about half of the solids in feces. By removing some of the bacteria from our intestines, the process prevents us from ballooning as these bacteria grow and reproduce in our bodies.[4]

"Water should be judged not by its history, but by its quality."[5] This statement by Dr. Lucas van Vuuren, an early pioneer of water reclamation in South Africa, is perhaps the most popular quote cited by professionals engaged in the world of potable reuse. Yet many people have difficulty ignoring recycled wastewater's past. Curiously, we're able to mentally override this kind of reaction in other matters. Brent Haddad, an economist at the University of California, uses the example of sleeping in hotel rooms: "There's a really good chance that that pillow was in contact and having experiences that would just be appalling to the next person who comes to the room," notes Haddad. But we tell ourselves that the cleaning crew came through and now everything is clean, and we're perfectly fine sleeping on that bed. "We frame out any history of that hotel room."[6]

As a result of our deep-seated aversion to human feces, potable reuse is a bonanza for psychologists, sociologists, and communication specialists who study perceptions of risk and how to improve public acceptance of recycled water. Among the phenomena of interest is the *magical law of contagion*—simply put, "once in contact, always in contact." In other words, anything that touches something disgusting becomes disgusting. While such "magical" properties have been ascribed to "primitive" people, they reflect a fundamental aspect of human reasoning.[7]

University of Pennsylvania psychologist Paul Rozin spent decades studying people's irrational feelings of disgust. In one experiment, Rozin asked his subjects if they would put on a sweater that Hitler had worn. "Almost everybody says 'no,'" said Rozin. Then he asked what if it was dyed to look completely different? Or say the yarn was unraveled and made into a new one? Would they put it on? Most people said no. There was one exception, however. If Mother Teresa put the sweater on first, some of Rozin's study subjects would consider putting it on too. In some way, her goodness would cancel out Hitler's evil.[8] Ironically, anyone drinking a glass of water in Europe is likely ingesting at least a few water molecules that at some point passed through Adolf Hitler's body—a truly disgusting thought.[9]

Like it or not, the term *toilet to tap* appears to be a permanent part of the water-reuse lexicon. It's almost irresistible as a headline for journalists who want to grab the reader's attention. Fortunately, the sensational headline is usually followed by more responsible information. Recognizing the inevitability of its continued use, communication specialists suggest recognizing the phrase and then moving on from there. "Toilet-to-Tap—Get Over It!" advised Patricia Tennyson, a prominent communications expert, during her talk at a national symposium where the U.S. Environmental Protection Agency (EPA) released its draft National Water Reuse Action Plan.[10] Sure enough, the first news story covering the plan was titled "EPA Water Reuse Plan Flush with Ideas Such as Toilet-to-Tap."[11]

Some water-reuse specialists have suggested that when *toilet to tap* come ups, just explain that it's really more like "toilet, treatment, treatment, treatment, aquifer/reservoir, treatment, treatment, tap." Wastewater utilities also have rebranded their facilities to emphasize that wastewater is a resource and not just waste to be disposed of. For example, Santa Barbara, California, renamed its El Estero wastewater treatment plant the El Estero Water Resource Center, with the tagline "Enhancing Santa Barbara's Quality of Life."[12]

For some people, no level of wastewater treatment or wordsmithing will ever make recycled water acceptable for drinking. Fortunately, most people can be convinced of the benefits of potable reuse. To get off to the right start, branding the product using some variation of "pure water" seems to be an effective way to get people to think more about the water's current quality and less about its past life. Winning messages also include emphasizing recycled water as good for the environment and as a locally controlled, drought-resistant water supply. Also helpful is providing people with a better understanding of the local water situation and why water reuse is beneficial, or even necessary. Ironically, blind taste tests have shown that people actually prefer the taste of recycled water over conventional tap water.[13]

Gaining public support for potable reuse takes time. It's also not simply a matter of the experts choosing a solution and then figuring out ways to convince everyone else. Elected officials, opinion leaders, as well as the general public need to be involved in a meaningful way— both early and often. *Legitimacy*, a cornerstone of a successful potable-reuse endeavor, depends on creating an "authentic conversation" about water reuse with the community at large. Ultimately, legitimacy can be gaged by how "taken for granted" water reuse becomes.[14]

While no Mother Teresa antidote exists for making potable reuse legitimate, there's no better way to start a conversation about potable reuse than over a beer. As Joanna Allhands with the *Arizona Republic* observes, "Hand people an ice-cold, crystal-clear cup of recycled water, and you're likely to get some upturned noses. But hand them a cloudy, pale yellow beer made from the stuff, and suddenly everyone's a recycled water connoisseur."[15] There's some irony here, given that the phrase *toilet to tap* had its origins with a beer company.

Beer makers are very picky about their water and actually prefer purified water. Normally, they have to remove minerals such as calcium and sodium and then re-add them to get the beer flavors they seek. With purified water, they start with a clean slate and can then add the desired ingredients at precise levels. The use of purified water for beer began in a surprising place.

Clean Water Services operates four wastewater treatment plants near Portland, Oregon. It's the largest water-reuse provider in the state, primarily for irrigating golf courses, parks, and athletic fields. Around 2014, the utility was thinking about expanding its reuse program. Art Larrance, considered "the godfather of craft brewing in Oregon," sat on their advisory council. "If you really want to talk about water, you've got to make beer," Larrance told the utility. "Beer starts conversations."[16]

The utility had to jump through quite a few hoops because Oregon regulations did not allow potable reuse. Through persistence, they managed to obtain permission to generate a small amount of recycled water for home brewing. In partnership with Oregon Brew Crew, the state's oldest home beer-brewing club, the first Pure Water Brew Sustainable Beer Challenge was held in 2014. Contestants were required to use recycled wastewater as the base for their brews. The recycled water was drawn from the Tualatin River, directly downstream from one of Clean Water Services' wastewater-treatment plants, and was treated using an advanced water-purification process much like that used in Orange County, California. That first batch contained 30 percent treated wastewater. The following year's competition upped it to 100 percent "sewage brewage."[17]

The idea quickly spread to other cities, including Boise, Denver, Milwaukee, San Diego, Scottsdale, Tampa, and as far away as Singapore. In 2017, Stone Brewing partnered with Pure Water San Diego to develop the limited edition "Full Circle" Pale Ale served at a local event the same week as World Water Day. The craft beer was described as "refreshingly clean! Brewed with very generous additions of Riwaka and Wai-iti hops, Pale 2 row malt, malted rye, and wheat malt. It also had tropical fruit notes from the New Zealand hops and some nice caramel flavors."[18] San Diego Mayor Kevin Faulconer, one of the first to enjoy a glass, declared it "fantastic."[19] The media had a field day. "This Brewery Is Making Beer Out of Poop Water. And It's 'Delicious'" went one headline. The Late Late Show had some helpful suggestions for the next batch: "Stone I, Pee, A," "Stone Extra Brown Ale," and "Stone Someone Had Asparagus Last Night Lager."[20]

In 2018, to celebrate Denver Water's 100th anniversary, Declaration Brewing Company teamed up with the Pure Water Colorado Demonstration Project to create the "Centurion Pilsner," the first commercially sold potable-reuse beer (as a limited edition). Colorado governor John Hickenlooper took a taste and declared, "It's delicious!"[21] Hickenlooper, a former brewpub owner, should know. But then again, the governor and former petroleum geologist also once drank fracking fluid, later declaring that it wasn't "tasty" but "I'm still alive."[22]

Another early adopter, Arizona Pure Water Brew (Pima County) hit the road with a mobile advanced water-purification truck that can produce about three gallons per minute of purified water. Clean Water Services followed with a thirty-two-foot-long Pure Water Wagon that has swing-up doors like a food stand at a fair.[23]

You can tell people that all water is recycled water and that they've been drinking recycled water all their lives, as most river water contains treated sewage from towns upstream. But that only gets you so far in persuading people to drink the stuff. They want more assurance that recycled water is "safe" to drink. Of course, no drinking water option (or virtually anything we do) is totally risk free. The fundamental question is, What is the *acceptable risk*?

Epidemiological and toxicological studies are one way of gauging risk. Three epidemiological studies of drinking water originating from the Whittier Narrows spreading basins in 1984, 1996, and 1999, as well as a few other locations, have not found any adverse health effects (cancer, mortality, or birth defects) from drinking recycled water. These studies are limited by the short periods to detect chronic disease and unknown exposures to recycled water of the population studied. They also can't prove there's no adverse effects, because it's impossible to prove a negative. Nonetheless, they didn't find any.[24]

Early toxicological studies in Denver and Tampa used rats and mice that were exposed to concentrates of reclaimed water. No adverse health effects turned up. These animal studies addressed a narrow range of potential adverse health effects. Despite these complications, the results, and other studies that followed, provide some evidence that risks to public health are low.

Another way to assess the safety of potable reuse is to compare it to conventional drinking-water supplies—a view promoted by the National Academy of Sciences in its first study of water reuse in 1982.[25] Following up on this idea, a 2012 National Academy of Sciences committee developed scenarios to compare the relative risk of de facto (unplanned) reuse with two potable-reuse scenarios—surface spreading and direct injection of treated wastewater. The two potable reuse scenarios assumed no dilution or blending with other waters. For de facto reuse, the committee assumed an annual wastewater content of 5 percent, a value that the committee viewed as reasonably commonplace.[26]

The committee concluded that the chemical risks of potable-reuse scenarios did not exceed the risk in common water supplies. They further concluded that the microbial risk from potable reuse "does not appear to be any higher, and may be orders of magnitude lower, than currently experienced in at least some current drinking water treatment systems" (i.e., from de facto reuse). Essentially, if everything goes along as planned, potable reuse is likely to be safer than conventional drinking water systems. It's the "as planned" part that requires attention.

As with anything else in life, unplanned events that affect water treatment can occur in multiple ways. An outbreak of disease or sudden toxic chemical load might overload the system. A treatment barrier can fail to operate properly because of clogged filters or torn membranes. A new contaminant may enter the system. And, of course, there's just plain human error. Unplanned events are not unique to potable reuse, but they do require greater vigilance. In particular, the higher loading of pathogens makes the consequences of failure potentially greater for potable reuse than for conventional drinking-water systems. Utilities address these risks through multiple means.

Just as with the 3Rs (reading, writing, and arithmetic), there are also 3Rs for all types of waste—reduce, reuse, and recycle. Potable-reuse specialists take it one step further to the 4Rs: reliability, redundancy, robustness, and resilience.[27] These concepts are lynchpins for achieving acceptable risk in water reuse.

The overarching "R" for potable reuse is *reliability*, which is the ability of potable reuse to consistently meet or exceed the public health protection provided by conventional drinking-water supplies. The word *consistently* is key and is supported by the other Rs.

Redundancy refers to the use of multiple barriers that exceed the minimum requirements of treating a particular contaminant. For example, an additional treatment barrier might be added to achieve a 15-log reduction in viruses under normal operating conditions when only a 12-log reduction is required by regulations. This ensures that if one of the methods isn't performing as designed, the system can still perform as intended.

Robustness focuses on the diversity of barriers. The idea is to proactively mitigate the next unknown chemical or pathogen by including a variety of treatments that are effective in different ways—such as biological treatment, adsorption, oxidation, and UV light. Robustness recognizes that no single process is effective against chemicals that run the gamut from large to small, charged to uncharged, strongly sorbent to weakly sorbent, and so forth.[28]

Redundancy and robustness are complementary ways of preventing failure and depend on the multiple-barrier approach of potable reuse.

Despite all efforts, however, failures can still occur. The fourth R, *resilience*, is the ability of a treatment train to successfully adapt to a failure in the system. For example, the system can be designed to automatically send untreated or undertreated flows to waste during power outages or when critical limits on treatment are violated. In these scenarios, failure occurs but does not affect public safety. Resilience becomes particularly crucial for direct potable reuse. It also brings us to the hazard analysis and critical control point (HACCP) framework that is designed to identify problems in the treatment processes in a timely way.

The HACCP framework was developed in the late 1950s to ensure adequate food quality for the nascent National Aeronautics and Space Administration. It was further developed by the Pillsbury Corporation and is now widely applied in the food and beverage industry.[29] The basic idea is to move beyond just testing the final product to determine quality. Instead, critical intermediary steps in the process are evaluated on an ongoing basis.

Despite its long name and awkward acronym, applying the HACCP framework to potable reuse is conceptually simple: Identify easily measurable physical or chemical properties that correlate with how well a treatment process is working. Measure those properties on an ongoing basis before and after that process. And finally, establish critical limits outside of which specified response actions will be undertaken.

Operators of water-treatment plants for potable reuse typically employ sensors that continuously monitor water-quality parameters, such as residual chlorine, turbidity, and conductivity, to alert operators of process upsets, fluctuations in the composition of incoming water, or changes in the performance of a treatment process. HACCP monitoring complements daily/weekly lab-analyzed monitoring, while also generating historical datasets on process performance.

Orange County is a good example. The broadest and simplest measure of the organic content of water is total organic carbon (TOC). This measurement is useful for determining how well reverse osmosis is screening out organic chemicals. Every few minutes, the Orange County Water District measures TOC in the inflow and outflow of RO to provide near real-time feedback to operators. If the TOC exceeds certain limits in the outflow, it triggers corrective action ranging from an investigation into the cause to shutting down the system. The TOC measurement is supplemented by semi-continuous measurements of the chlorine content, conductivity, and turbidity of the RO inflow and outflow.[30]

As another example, an extremely small amount of short-circuiting in the UV system can result in some pockets of water receiving a lower dose and viruses slipping through. To detect these sorts of problems, the Orange County Water District measures UV dose, power, and transmittance with semi-continuous online analyzers.

Muriel Watson of the Revolting Grandmas came away from her tour of the Orange County plant unimpressed. "It's not the sun and the sky and a roaring river crashing into rocks"—nature's way of purifying water. "It's just equipment," she said.[31] Watson missed an important ingredient beyond just equipment—the human dimension of this complex technology.

A tour of Orange County's Groundwater Replenishment System led us to a control room where operators are constantly monitoring a computer system (commonly known as SCADA for "supervisory control and data acquisition") to track processes throughout the plant on a real-time basis. John Sonza, a plant operator, and Randy English, who operates the injection wells, have more than forty years of experience and are clearly dedicated to the importance of their job. Such operators require special training and certification on advanced water treatment, emergency-response procedures, and drinking-water regulations.

A disease outbreak in Walkerton, Ontario, in 2000 is a cautionary tale of what can go wrong with lax operation. The outbreak occurred after a heavy spring rainfall that resulted in one of the city's shallow wells becoming contaminated by pathogens from manure on an adjacent farm. A virulent strain of E. coli killed seven people, with more than two thousand others becoming seriously ill. All told, the bacteria affected almost half the population of this small farming community, located about one hundred miles northwest of Toronto.[32]

Contamination most likely entered the well on May 12, a week before illness became evident in the community. When asked on May 19 (and again on May 20) whether there were any problems with the drinking-water quality, the general manager of the system assured local health authorities that the water was satisfactory—despite having received adverse microbiological monitoring results a couple days earlier. Practices at the facility were sloppy. The system supervisor was supposed to measure the chlorine residual daily but failed to do so on most days and recorded fictitious entries on the daily operating sheets.

On May 21, the Ontario regional medical officer and his staff began marking a town map with a yellow dot for each diarrhea case. By the end of the day, the map was yellow. "We knew there was only one thing that can do that—the water supply," he reported.[33] Health officials immediately issued a boil-water advisory, but it was too late. The first victim died the next day. At least eight days without valid chlorine residual monitoring had passed between the well contamination and the boil-water advisory.

Failures occurred at many levels, including poor system management and operations, inadequate operator training, inadequate watershed protection, and ineffective regulatory oversight by the Ontario Province. The operators who falsified records had no idea of the risks that they were bringing on their community, as they continued to drink the water themselves during the outbreak. Not knowing the source of contaminants, the local hospital made the situation worse by recommending that parents have their children with diarrhea drink more fluids, thereby increasing their exposure to the contaminated water. The moral of the story is that it's not just the equipment but the operators that matter.

More recently, a 2021 cyberattack on the drinking water plant of Oldsmar, a small town near Tampa, Florida, illustrated the importance of trained and vigilant operators together with built-in resilience of the treatment systems. The hacker used a remote-access program shared by plant workers to briefly increase the amount of lye (sodium hydroxide) by a factor of 100. The attempt was easily thwarted by an alert operator but is a reminder that municipal water utilities are a soft target for cyberattacks.[34]

Development of innovative monitoring sensors, advanced analytics, and artificial intelligence to enhance the safety and reliability of treatment systems is an active area of research. These emerging technologies can supplement, but not replace, trained and dedicated plant operators.

Nonpotable use is also not fail-safe, as there have been sporadic incidences of cross-connecting purple pipes with potable water. The Otay Water District in Chula Vista south of San Diego is a purple-pipe pioneer and primary customer of San Diego's nonpotable recycled water.

This water district also produces its own recycled water, dating back to the late 1960s with a small recycling facility dubbed "Miss Stinky."[35]

In 2007, it was discovered that people at seventeen businesses in a Chula Vista office park had been drinking nonpotable water for about two years because of accidental cross-connection of purple pipes with drinking water. A mix of four parts drinking water, one part nonpotable water came from their taps. It wasn't until water deliveries to the buildings increased to 100 percent nonpotable water that the problem finally came to light. The water not only tasted bad; it had a yellowish tint. "You would flush the toilet, and it looked like it wasn't flushed," complained one owner.[36] People suffered from frequent gastrointestinal illnesses without knowing why. Once the problem was discovered, the water district was held liable for several million dollars in damages to the tenants affected by the faulty hookup and to the owner of the business park, who couldn't find businesses willing to pay the going rental rate after the incident.[37] Other isolated incidents from cross-connections have occurred, but no major public health problems have been identified in the United States from using dual distribution systems.[38]

Nonpotable reuse is also not immune from controversy about its safety in other ways, as illustrated by Redwood City, California.[39] Located midway between San Francisco and San Jose, the city is dependent on imported water from San Francisco's Hetch Hetchy Reservoir in the Sierras for virtually all its water. By the end of the 1990s, Redwood City was exceeding its contractual limit and dipping into water underutilized by other communities. This opportunity was coming to a close as the needs of the communities dependent on the water supply grew. With few options, Redwood City's consultants proposed constructing a nonpotable-reuse system, along with increased conservation measures.

After a brief pilot project, the city determined to move ahead with nonpotable reuse for Redwood Shores (the Shores), a waterfront community of about five thousand homes on San Francisco Bay. With a wastewater-treatment plant nearby, the Shores neighborhood offered good potential for potable-water savings. The rest of Redwood City would follow later.

The first public outreach occurred in June 2002, when a workshop was held in the Shores to inform the community of the plans. The workshop was advertised in the local Shores newsletter and a regional newspaper. Only two people showed up, but they would end up having a major impact on the public's perception of the project. City representatives were peppered with questions for hours.

When one of the two attendees asked if she could opt out of the program, she claimed to be laughed at and told that all residents would be required to use water from the system to water their yards.[40] Mandatory ordinances on water projects are common to qualify for government funding, but the two attendees felt that the project was being foisted on them without their consent. They formed an opposition group known as the Safewater Coalition.

Controversies included concerns about rising water rates and whether the project was a ploy for more development, but the main focus centered on the possibility of children ingesting contaminated water from sprinklers at homes, parks, and schools. A second forum was held in September 2002 to address these issues. Thirteen panelists were selected by both the city and the Coalition. This time about a hundred people showed up.

The public pressure continued, and in February 2003, the city council backed off the mandatory-use requirement—recycled water would be optional for existing residences and homeowners' associations. Opponents continued to be concerned about possible health risks to children from using recycled water in parks, playgrounds, and schoolyards. "Our grass can go brown so long [as] the kids are safe," emphasized one of the Safewater Coalition founders.[41]

In July 2003, after a marathon eight-and-a-half-hour meeting, the city council made a wise decision. Faced with continuing opposition that threatened the project's successful implementation and wanting to heal a divided community, the council decided to give the project's opponents an opportunity for working with the city to find a mutually agreeable recycled-water project solution.[42]

The application process for the Redwood City Recycled Water Task Force was open to the public, with the intent to form a balanced group. Of the twenty people selected, nine were for recycled-water use, nine were opposed, and two were neutral. One of San Diego's Revolting Grandmas, who seemed to be everywhere that recycled water was being discussed, got on the task force.

Within six months, the task force arrived at a compromise solution that cost about the same as the original plan. The dual distribution system would only deliver the recycled water to places where there was limited potential for human contact, thereby avoiding schools, parks, and playgrounds. To compensate for less recycled water, the task force recommended a number of water-conservation measures, such as installing artificial turf on the city's sports field. They also identified some minor groundwater use. In March 2004, the city council enthusiastically accepted the task force's recommendations. The Shores project went online in 2007, and Redwood City continues to extend nonpotable water to other parts of the city.[43]

The take-away lesson from Redwood City is the need to engage the public early—even on projects thought to be noncontroversial. It also demonstrates the power of a small but determined opposition group. In Water 4.0, David Sedlak notes that the chances of getting sick from the pathogens present in a few drops of water from a sprinkler that is hooked into a water reuse system is hundreds of times lower than other pathogen risks that we readily assume, such as visiting a petting zoo, swimming at a beach where stormwater runoff has contaminated the sand with bacteria, or eating fresh organic produce from a farm that uses animal manure for fertilizer. But the risks are never going to be zero. The key is achieving acceptable risk.[44]

The most common concern associated with nonpotable reuse is the potential transmission of infectious disease from microbial pathogens by inadvertent ingestion of recycled water, skin contact, or inhalation of aerosols. Thus, regulations for nonpotable reuse focus mainly on mitigating health risks from microbial pathogens through disinfection and/or by imposing use-area controls (e.g., fencing, signage, and buffer zones).[45] How the messaging is stated can be important. For example, a simple change in signage around areas where nonpotable water is used for irrigation from "Do not drink" to "Not for drinking" changes the overall public perception of water reuse, while achieving the same end result.

There are striking similarities between the Redwood City controversies and another high-profile case a few years later in Toowoomba, Australia. In 2005, this orderly city known as "Queensland's Garden City," was in the fifth year of the Millennium Drought, the worst drought in the country's history. The city's three reservoirs were rapidly being depleted. Behind the scenes, the city was working on an indirect potable-reuse project using treatment technology similar to Orange County, California. The advanced-treated water would be added to one of the city's reservoirs.

The first public discussion of this solution to Toowoomba's water problems came in May 2005, when the city's mayor, Dianne Thorley, enthusiastically unveiled the idea at the monthly meeting of the women's club. "The ladies in that room were dumbfounded," recalled Rosemary Morley, past president of the Chamber of Commerce and major opponent of the project.[46] Morley was particularly aggravated by the mayor's insistence that the decision to proceed was a done deal. "No consultation, no debate," she later said. "That's like waving a red flag in front of a bull."[47] Morley formed an opposition group called CADS—Citizens Against Drinking Sewage (or "Citizens Against Drinking Shit," depending on the audience). Clive Berghofer, a cantankerous former Toowoomba mayor and millionaire property developer, soon came out as a powerful force against the recycling project. Concerns included effects on residents' health and the city's image and attractiveness to business.

The project's proponents were generally outgunned and outmaneuvered by the opposition. Being the first to communicate with the public, CADS had the "First Mover Advantage," and became the main source of information. By February 2006, ten thousand people had signed the CADS petition against the potable recycled-water initiative.[48] Against the city council's wishes, which wanted to undertake a three-year community-engagement program, the commonwealth held a referendum on the project. Despite the intensifying drought, with dam levels at approximately 20 percent, 62 percent of the voters turned down the water-recycling plan.

Toowoomba's water supply was supplemented by emergency construction of a twenty-four-mile pipeline from Lake Wivenhoe, one of the Brisbane area's reservoirs. This came with an ironic twist. A few years after the Toowoomba initiative was defeated, advanced wastewater-treatment plants were constructed in Brisbane for indirect potable reuse. This Western Corridor Recycled Water Scheme is not currently in use, but during severe droughts, purified wastewater from the project could be discharged to Lake Wivenhoe. In other words, Toowoomba residents could end up drinking recycled wastewater after all. However, the opposition has diminished. A survey taken two years after the Toowoomba referendum defeat revealed much less resistance in the community to drinking recycled water. Knowing that they were not singled out and that Brisbane also would be drinking recycled water may have allayed some concerns.[49]

A more successful example of water reuse is the tiny island city-state of Singapore, home to almost six million people. Singapore is recognized worldwide for its holistic approach to water management, making use of "Four National Taps."[50] The first is imported water from Malaysia. Having little in the way of natural water resources despite a wet climate, Singapore has long been dependent on Malaysia for much of its water supply. The other three National Taps—local runoff, desalinated seawater, and reclaimed water—are motivated in large part by a desire to free Singapore from this dependence on a country with which it has a rocky relationship. In Singapore, water is viewed as a national security issue.

Singapore captures runoff from two thirds of its land surface through a network of about five thousand miles of waterways and seventeen reservoirs. These are much more than simply engineering structures. Driven by a vision of sparkling rivers with landscaped banks flowing into picturesque lakes, Singapore has been transformed into a City of Gardens and Water. To make the runoff as clean as possible, the city-state rigorously enforces laws related to land use, automobile maintenance, and application of chemicals on buildings and gardens.[51]

The third National Tap, desalination of seawater, helps meet about 30 percent of Singapore's water demand but is energy intensive. This leads to the desirability of its fourth National Tap. Known as NEWater, advanced-treated water meets up to 40 percent of the nation's current water needs, with a goal to increase this to more than half of Singapore's water supply by 2061—when the current agreement with Malaysia ends. Most of the reclaimed water is supplied directly to industries, including those that need very high-quality water for wafer fabrication and electronics manufacturing. A small percentage is discharged to reservoirs for indirect potable reuse. Singapore is widely known for the sophistication of its water-quality monitoring, including real-time monitoring of both raw and treated water. Singapore's NEWater program is also lauded for its public-relations success, including a popular visitor center for citizens to learn firsthand about the program.

Chapter Twelve

Serving It Straight Up

With potable reuse, it's like it rains every day.

—Mayor Stephen Santellana, City of Wichita Falls, Texas[1]

The first direct potable reuse by the United States received no serious objections by its end users. It also solved a major challenge: How to more effectively secure drinking water for the astronauts at the International Space Station.

Shipping water to the space station costs anywhere from $10,000 to $90,000 per pound.[2] That translates to somewhere between $44,000 and just under $400,000 to ship two liters of water—the recommended amount a person should drink each day to stay hydrated. Water reuse is an obvious solution to cut costs. The ability to recycle water is also key to deeper exploration of space by humans.

In the early phases of the International Space Station, astronauts supplemented their water supply using a system adopted from the Russian Mir space station. Condensate from humidity was collected and processed into potable water.[3] NASA's water-recovery system, launched in 2008, took it a step further by also recovering urine.

A urine processor boils astronauts' urine for treatment by distillation. Water produced by the urine processor is combined with the other wastewaters and sent through a series of filtering materials and chemical reactions for purification. The produced water is tested by onboard sensors; unacceptable water is cycled back through the water-processor assembly until it meets purity standards. Clean water is sent to a storage tank and is ready for the crew to use. As a running joke, the astronauts quip: "Yesterday's coffee is tomorrow's coffee."[4]

The recycled water is also useful to help with another basic part of the life-support system—oxygen to breathe. An oxygen-generation system uses electrolysis to separate water into hydrogen and oxygen. The solar panels on the space station easily provide the electricity needed for this energy-intensive process. The hydrogen is discarded into space, and the oxygen is used, in NASA lingo, for atmosphere revitalization.

Direct potable reuse at the International Space Station is not quite a fair comparison to its use on Earth. Here, potable reuse deals with a much more complicated waste stream that includes feces and a plethora of chemicals.

Direct potable reuse (DPR) introduces highly treated wastewater either directly into a public water system or into the raw water supply immediately upstream of the intake of a drinking-water plant. By treating water at a location that is readily connected to the water-supply distribution system, DPR has the potential for higher water recovery, lower treatment costs, and less infrastructure for water transport than indirect potable reuse.[5] It also avoids going through all the trouble and expense of advanced treatment to obtain purified water, and then putting it back into the environment where there's a possibility for water-quality degradation. In some cases, DPR may be the only option for potable reuse in the absence of a suitable nearby groundwater basin or surface reservoir.

The downsides of DPR are the advantages that arise from environmental buffers. Groundwater or surface-water storage allow time for mixing, dilution, and natural processes to improve the quality of the recycled water. DPR can make up for the loss of this supplemental natural cleansing by more rigorous treatment processes. The biggest challenge with DPR is the extremely narrow window of time to take action in the event of treatment failures. An engineered storage facility, however, can provide some response time for verification of specific water-quality parameters before the water enters the distribution system.[6] With indirect potable reuse, reintroduction of the purified water into the environment potentially allows the public to eliminate or reduce the mental association with the water's wastewater origin. The extent of this psychological advantage over DPR is unclear.

Scientific views on the viability of DPR have evolved considerably over time. A 1998 National Research Council report minced no words: "Direct use of reclaimed wastewater for human consumption, without the added protection provided by storage in the environment, is not currently a viable option for public water supplies."[7] More recently, experts have suggested that improvements in treatment and monitoring technologies (perhaps including an engineered storage buffer) can replace the need for environmental buffers. A 2012 study by the National Research Council recognized the important roles that environmental buffers can play in ensuring public acceptance of potable water reuse but concluded that "the historical distinction between direct and indirect water reuse is not meaningful to the assessment of the quality of water delivered to consumers."[8]

In 2018, Swiss and U.S. researchers offered a more cautionary assessment of DPR. According to the researchers, insufficient attention is being given to catastrophic risks with low probabilities of occurrence, but high consequences. They noted that such events have emerged in other seemingly "fail-safe" systems with disastrous results. The *Titanic*, Fukushima, and Deepwater Horizon are well-known examples.[9]

These concerns include "Black Swan" events, a concept brought to people's attention in a popular book by author Nassim Taleb.[10] A Black Swan event has three characteristics. First, it lies outside the realm of regular expectations, because nothing in the past can convincingly point to its possibility. Second, it carries an extreme impact. Third, in spite of its outlier status, humans can concoct explanations for its occurrence *after the fact*. To summarize the triplet: These events are outliers, have extreme impacts, and are predictable in hindsight but not in foresight. The rise of Hitler was a Black Swan event, as was the election of Donald Trump and the response to COVID-19. For DPR systems, a Black Swan event might arise from a pandemic of a new pathogen or a large spill of dangerous chemicals into the sewage system. Or simply from operator failure or treatment-system failure or some combination of the two. Of course, it's important to keep in mind that Black Swan events also can occur in conventional drinking-water systems. But the stakes are higher with DPR.

Low-probability, high-consequence events for direct potable reuse carry risk not only to public health but also to the industry as a whole. A major system failure in a direct potable-reuse plant could have extensive negative spillover effects, including an irreversible loss of public trust in the technology. Particularly in the emergent phase of a new industry, a catastrophic system failure can delegitimize a technology for an extended period of time.[11]

The risks of low-probability, high-consequence events can never be fully avoided, but lessons from other industries provide insights on how to improve the safety net. The researchers pointed to the history of risk management by three other industries: aviation, offshore oil drilling, and nuclear energy. Regulators in these industries have learned the hard way that, although catastrophic system failures can never be completely ruled out, their probability and impacts can be significantly reduced by establishing and nurturing an industry-wide safety culture, as well as by creating an independent auditing and self-policing organization. The latter assures that any system failures or "near misses" are investigated and adequately addressed.[12]

Direct potable reuse ups the ante for safety. The 4Rs previously discussed, strict source control, and the ability to rapidly divert advanced-treated water that does not meet specifications become particularly important with DPR, as does operator training and certification. Critical treatment processes should be monitored daily, and online metering of surrogate parameters (e.g., electrical conductivity before and after reverse osmosis) is needed to monitor treatment performance in near real time.[13]

Lying just above South Africa, Namibia is the driest country in sub-Saharan Africa. The only perennial rivers are on the country's borders. One of earth's most astonishing water-harvesting adaptations is found here in the hyper-arid Namib Desert. When the occasional fog circulates over the Namib Desert beetle's back, tiny water droplets accumulate on top of bumps on its armor-like shell. These water-attracting bumps are surrounded by waxy water-repelling channels. When the droplets become large and heavy enough, they roll down these channels to a spot on the beetle's back that leads directly to its mouth.[14] Scientists are likewise trying to develop passive, low-cost fog-harvesting approaches, particularly for places without access to safe drinking water.[15]

Windhoek, the capital and largest city of Namibia, is situated on a high central plateau between the Namib and Kalahari Deserts. Like Namibia's desert beetle, Windhoek has developed a special adaptation for obtaining its drinking water. In 1968, it became the first city in the world to return treated sewage effluent directly to its potable water system.[16] They began modestly, averaging around 4 percent of their water for many years. After several treatment-plant upgrades, recycled water grew to 35 percent of the drinking-water supply during normal periods, and as much as 50 percent during droughts. Research has found no evidence of harmful health impacts, and initial public opposition faded over time.[17]

Although not recognized as DPR at the time, in many respects the first U.S. case of direct potable reuse took place in a small town in eastern Kansas in response to a short-term emergency. Chanute, Kansas, obtains its drinking water from the Neosho River, a tributary of the Arkansas River. During the summer of 1956, the Neosho became the "No Show" river when it ceased to flow in the midst of a severe drought. Desperate for water, the city dammed the river below the outfall from the wastewater-treatment plant, thereby backing up secondary-treated wastewater to the intake for the town's drinking-water plant. The seventeen-day retention time in the makeshift reservoir provided some water-quality improvement but was effectively direct potable reuse.[18] No known adverse health effects occurred during this five-month emergency period, but it was hardly a rip-roaring success. The recycled water was pale yellow with an unpleasant odor and musty taste. It also foamed when agitated. Bottled water sales flourished.[19]

Big Spring, Texas, was the first town in the United States to undertake planned direct potable reuse. This small West Texas town may seem like an unlikely place to be using treated sewage effluent in its drinking-water supply, but they don't have much choice. The town's "big spring," originally fed by a small but prolific aquifer, dried up almost a century ago after the arrival of a railroad and the West Texas oil boom.[20] By 2002, the Colorado River Municipal Water District (CRMWD), which provides water to Big Spring and nearby cities, had fully tapped the area's surface-water and fresh groundwater resources. To keep up with the needs of the growing region, the CRMWD considered desalinating brackish groundwater but found DPR to be less expensive.[21] With over sixty inches of evaporation a year, indirect potable reuse wasn't considered an option.

The CRMWD held public meetings, appeared on television and radio, and gave talks to civic clubs. The idea was well received by the public. At one meeting, a man joked that he thought the idea was great because he would get to drink his beer twice.[22] In actuality, he probably hadn't been drinking the water at all. For many years, people had complained about the taste. "Nobody drinks the water here," said one resident. Another responded, "Hell no! We don't do that. I'll bathe in it, but I won't drink it. It's too hard. . . . It's nasty."[23]

In 2013, the CRMWD began operating the nation's first (earth-based) planned DPR facility. After secondary wastewater treatment, the water is treated using a combination of microfiltration, RO, and ultraviolet disinfection, then blended with reservoir water, and finally distributed to five drinking-water facilities, where it is treated again.[24] While the plans for the advanced-treatment plant had been made during a non-drought period, it came online none too soon during an intense drought. By January 2014, the closest reservoir was about 1 percent full.[25]

The state of Texas evaluates treatment requirements for DPR on a case-by-case basis. Pathogen treatment requirements (i.e., LRV or log-reduction values) are determined by evaluating the pathogen loads in the specific wastewater effluent that is to be used for DPR. These individualized treatment requirements may change over time if warranted by ongoing monitoring programs. This regulatory approach has allowed the state to adapt its approach to several different DPR scenarios.[26]

Wichita Falls, Texas, a city of over one hundred thousand lying near the Oklahoma border, followed in Big Spring's footsteps to avoid running out of water during the worst drought on record. This sunbaked town, which hosts the Hotter'N Hell Hundred bike ride each August, is not afraid of challenges. The intense bike ride was first held in 1982—one hundred miles in 100-degree heat to celebrate the city's centennial. A New York marketing firm had proposed a rocking chair marathon. They were promptly fired. The bike race was the brainchild of the local postmaster and seemed like a perfect fit for the town's pioneering spirit. Thousands of people come to compete in the grueling event every year. Among its claims to fame, Michael Eidson invented the popular CamelBak hydration pack, repurposing an IV bag in a tube sock, while competing in the race.[27]

In 2011, Wichita Falls had one hundred days over 100°F, earning it the Weather Channel's title of "#1 Worst Summer Anywhere in 2011." City reservoirs were drying up, and groundwater wasn't available for backup—the nearest high-quality groundwater is the High Plains aquifer over two hundred miles to the west. If 2011 conditions continued, Wichita Falls was projected to be out of water within two years. The city approached the Texas Commission on Environmental Quality (TCEQ) about implementing a DPR project but only received permission for indirect potable reuse. As it became increasing clear that the city could run out of water before indirect potable reuse came online, TCEQ granted permission for the city utility to implement DPR on an emergency/temporary basis.[28]

The city had a leg up on potable reuse, having already installed microfiltration and reverse osmosis to treat brackish lake water. All that was needed was a twelve-mile aboveground pipeline to connect effluent from the wastewater-treatment plant to the advanced treatment system. But there were bigger challenges, including tests to demonstrate compliance with drinking-water regulations and pathogen-removal targets, development of monitoring plans, establishment of alarms and shutdown triggers, and operator cross-training between the wastewater and drinking-water plants. All of this occurred on a highly compressed schedule. To demonstrate that the milestones were met, almost two million discrete data points were collected on water quality over about a year of testing full-scale operation. The system came online in July 2014, blending a 50/50 mix of treated effluent and lake water. When a historic flood broke the drought a year later, the DPR project came to an end.[29]

From the outset, Wichita Falls officials made public communication and outreach a priority. Before presenting DPR to the public, officials worked with the media to create a video in which local doctors and university professors explained the project. A media blitz provided daily featured stories and updates on drought status. As in the case of Big Spring, no serious public opposition arose. During the use of DPR, the city received no taste or odor complaints. Ironically, complaints by customers came rolling in *after* the city stopped using DPR and the water went back to its previous salty condition. The public cheered when a permanent indirect potable-reuse project began operation in 2018, using one of the city's two freshwater lakes as the environmental buffer.[30]

El Paso, Texas, sits in the middle of the Chihuahuan Desert with an average annual rainfall of only nine inches. In the late 1970s, El Paso residents received some disturbing news. According to a modeling study, the city's primary drinking-water aquifer would be depleted by 2030.[31] This alarming announcement about the source of more than half the city's water supply attracted considerable media attention. Later projections moved the day of reckoning up to 2020.[32]

The predicted rates of groundwater depletion never materialized, in large measure because El Paso residents dramatically modified their water-consumption habits. But the die was cast. It was the beginning of "a lot of dialogue between the utility and customers," recalled John Balliew, the utility's CEO.[33] With El Paso also running out of areas to dispose of wastewater effluent, someone floated the idea to treat wastewater to drinking-water standards and inject it into the aquifer.

In 1985, El Paso Water opened the Fred Hervey Plant and began the nation's first well injection of treated wastewater specifically intended to augment an aquifer for water supply.[34] (Orange County, California, began well injection of treated wastewater nine years earlier, but its purpose was to create a seawater barrier.) The treatment process includes granular activated carbon and ozonation as barriers against chemical contaminants and pathogens. By avoiding the use of RO, the facility does not produce a brine waste that requires disposal.

Advanced-treated water from the Fred Hervey Plant was injected several hundred feet underground, where it co-mingled with fresh groundwater. It takes about six years for the water to filter through the ground before being pumped back out from downgradient wells, where it is treated and piped to El Paso residents. The treated wastewater undergoes considerable mixing with the native groundwater, with reclaimed water accounting for approximately 1 percent of the water withdrawn in the nearest downgradient wells.[35] The recycled water has helped stabilize aquifer levels with minimal controversy. In recent years, the utility has switched to spreading basins, which are more economical and easier to maintain than the injection wells.

Over thirty billion gallons of reclaimed water have been recharged into the aquifer.[36] Nonetheless, El Paso's water challenges continue in large part because of a diminishing supply from the Rio Grande. The "great river" begins high in the mountains of southern Colorado and flows more than 1,800 miles through New Mexico and Texas before emptying into the Gulf of Mexico. El Paso lies at about the midway point. In recent decades, climate change has reduced snowmelt runoff from Colorado and northern New Mexico into the Rio Grande. These reduced flows and higher evaporation rates have decreased storage in Elephant Butte Reservoir, which stores water for irrigation and municipal use in southern New Mexico and the El Paso area. At times in recent years, Elephant Butte Reservoir has been only a few percent full.[37] Irrigation in New Mexico below the dam has further reduced the flow of the Rio Grande at El Paso and is the subject of an ongoing U.S. Supreme Court case between Texas and New Mexico.

To bolster and diversify its water supply, El Paso turned to its underlying brackish groundwater. In 2002, with growth of the Fort Bliss Army post in El Paso constrained by its limited supply of well water, the U.S. government and El Paso Water agreed to jointly construct a desalination plant on base property. "We were happy, they were happy," recalled Ed Archuleta, El Paso Water's long-time CEO.[38] The world's largest inland desal plant opened in 2007.

The desalination plant, however, is not a silver bullet. El Paso's long-term water challenges with population growth and climate change continue. A severe drought in 2010–2014 helped motivate El Paso Water's latest ventures—banking Rio Grande water in groundwater and developing direct potable reuse. The utility also purchased groundwater rights from ninety miles away in Dell City, Texas, for future use perhaps fifty years from now.

El Paso's water rights to the Rio Grande exceed overall water demand, but the river is subject to large seasonal and annual swings in flow. In 2013, Rio Grande water was available for less than two months and barely covered the city's needs.[39] To gain independence from the vagaries of the river, El Paso Water plans to capture excess Rio Grande water during high-flow periods, treat it to drinking-water standards, and recharge it in arroyos that have been excavated down to sand. The arroyos will also replace the spreading basins for recharging treated effluent from the Fred Hervey plant. As a major plus, they will be constructed to be attractive and environmentally beneficial with wetland habitat and hiking and biking trails.

On top of these steps, El Paso is on track to become the first large city in the Northern Hemisphere to treat its wastewater and send it directly back into its taps. At ten million gallons per day, the plant would produce about 80 percent more potable water than the Windhoek plant. It would use a treatment process similar to the fully advanced treatment at Orange County, with the addition of granular activated carbon at the end.

Reverse osmosis is not part of the treatment train for aquifer recharge, but it will be for the DPR project. Leftover salt concentrate will have to be disposed of. To help address this problem, the Center for Inland Desalination Systems at the University of Texas at El Paso (among others worldwide) is developing processes to provide greater recovery of water and less brine from RO systems.[40]

El Pasos citizens have been remarkably receptive to the DPR idea. An early survey found that 84 percent of residents strongly approved of the project.[41] El Paso's history with safely implementing other forms of water reuse and desalination built trust with the community, as well as a familiarity with water from alternative sources. El Pasoans are also keenly aware that they live in the desert and can't take their water supply for granted. For many years, El Paso Water's popular mascot, Willie the Waterdrop, has visited schools, educating kids about the importance of water and the need for conservation. In 2008, El Paso Water opened the TecH2O Water Resources Learning Center, offering a museum-like experience for students with interactive exhibits.[42] The utility's efforts have paid off. Since the 1980s, per person water use has dropped by 35 percent.[43]

A multi-faceted, bilingual outreach program for the DPR plans began in 2013 and continues to this day.[44] Christina Montoya, communications and marketing manager for El Paso Water, notes that Gilbert Trejo, chief technical officer of El Paso Water, grew up in El Paso and not only is fluent in Spanish but also speaks it the way El Pasoans do. This gives him extra credibility with the public. Montoya also emphasizes the importance of having utility employees out in the community talking about the project. Not only do the employees have direct knowledge of the undertaking, but they can also talk in everyday nontechnical language. This approach also enhances teamwork in reuse among the utility employees.[45]

In 2016, El Paso Water piloted the advanced water-purification process. The utility offered tours of the pilot facility and provided speakers for clubs, schools, and businesses to explain the project and the treatment process. The approval rate jumped to as high as 96 percent after people had toured the pilot facility.[46] El Paso Water also courted the local media about the need for the DPR facility and how it would work. "They were kind of relentless in getting us to cover it," recalled a water-issues reporter for the *El Paso Times*.[47]

The Texas Commission on Environmental Quality gave El Paso Water approval to proceed with design of the full-scale facility. Construction is expected to start around 2024. By 2030, El Paso Water projects that about 6 percent of its water supply will come from direct potable reuse.[48]

About a hundred miles northeast of El Paso, the Village of Cloudcroft, New Mexico, is seeking to become the first community outside Texas to implement planned DPR. Located at 8,600 feet elevation near the top of the Sacramento Mountains, this resort town is named for the clouds that often blanket it. The town's permanent population is about seven hundred but often triples during the ski season, summer, and holidays. The quaint pioneer village has several claims to fame in addition to its efforts at DPR. The Lodge at Cloudcroft has hosted an eclectic mix of guests that includes Judy Garland, Clark Gable, and Pancho Villa.[49] For many years, Cloudcroft was also home to the highest golf course in North America.

Cloudcroft has historically relied on water from springs and wells, but drought conditions have severely challenged these water sources in recent decades. Many of the town's wells are in shallow alluvium and vulnerable to droughts. In 2004, a severe drought led to the National Guard trucking twenty thousand gallons of water daily up the mountain.[50] Since then, the town has experienced several additional water supply emergencies.

The village was not only periodically running out of water but also running out of options to address these shortages. Water conservation had already substantially reduced per-capita water demand, and Cloudcroft's location near the top of the mountains made it difficult to find new water sources. Several exploratory wells failed to find a suitable groundwater supply. Piping in water from outside the village boundaries is not feasible due to economics and water-rights challenges.[51]

Mayor David Venable ("Mayor Dave") wondered if it was possible to use the town's wastewater to solve the shortage. He took the question to Eddie Livingston, a water-resources consultant. They soon envisioned a direct potable-reuse project that would blend purified wastewater with spring and well water using an innovative treatment approach that works as follows: The existing wastewater treatment plant would be upgraded to a membrane bioreactor (MBR). An MBR combines a membrane process, such as microfiltration, with a biological process such as activated sludge. The permeate from the MBR would flow downhill by gravity several miles to the Water Purification Facility located near the town's wellfields. The pressure resulting from gravity flow through the pipeline would force the MBR permeate through a RO system, making the process very energy efficient. Advanced oxidation using ultraviolet light and hydrogen peroxide would follow. The water would be chlorinated and discharged into a one-million-gallon covered engineered storage tank. About ten days' storage time in the tank would provide time for corrective action. If monitoring of surrogate measurements suggested that the advanced wastewater-treatment facility is not attaining target pathogen reductions, it would be shut down immediately, and water would not be sent to the engineered storage buffer.[52]

The purified wastewater would be blended about 50/50 with existing spring and well water, potentially doubling the water supply. The blended water would be treated using ultrafiltration, UV disinfection, granular activated carbon, and final chlorine disinfection. After this treatment, the product water would be introduced into the village's water-distribution system.[53]

Disposal of RO concentrate in the inland area presents an additional challenge. Plans are to put the concentrate to beneficial uses such as dust control and storage for firefighting. It might also be disposed of using deep-well injection, possibly in one of the poor-water-quality wells dug when they were looking for additional sources of water.

To get started, Mayor Dave was able to tap into a water-innovation fund established by the governor. Construction of the DPR facilities began in 2009.[54] Because there are no potable-reuse regulations in New Mexico, the New Mexico Environment Department brought on the National Water Research Institute (NWRI) for regulatory assistance. An expert panel assembled by the NWRI concluded that the treatment train for the DPR project is sufficient to meet public health criteria, but also made recommendations for more robust water-quality monitoring. The panel's report was for Cloudcroft only but will be helpful in developing regulatory guidance for future DPR projects in New Mexico.[55]

The DPR project was nearly 85 percent complete when faulty construction issues brought it to a halt. By the time construction resumed, the treatment technologies had changed, requiring process retrofits and new equipment.[56] Fortunately, the town was able to complete the MBR in 2017, so at least the community now has a modern wastewater-treatment plant.

In 2019, the legislature appropriated funds for completion of the facilities. But meanwhile, the state requires an entirely new review and approval process by both construction and drinking-water bureaus in the New Mexico Environment Department. Complicating matters, a huge turnover in staff has resulted in a near total loss of institutional memory. Eddie Livingston submitted final design drawings for the advanced wastewater-treatment plant in spring 2020.[57]

One problem the project has *not* had is organized public opposition. Given the town's lack of options to address its exceptional water scarcity, public acceptance was fairly easily obtained. The village administration held three public meetings where water-supply options were discussed, but no formal education and outreach campaigns were undertaken. The only significant concerns were the cost and possibility of a rate increase.[58]

Because the project has received grant funding to build the DPR facility, so far Cloudcroft has not needed to raise water rates. Once up and running, however, operational and maintenance (O&M) costs will likely become a key issue to cover salaries for highly trained staff, as well as power, chemical, and monitoring costs. The community's already-high water rates could potentially double.[59] As a tourist mecca, however, the town is fortunate in having the possibility of an excise tax on hotel rooms and restaurants to help pay for the O&M costs.

Cloudcroft is not alone among small to medium-sized communities facing exceptional water challenges. A 2005 report by the U.S. Department of the Interior predicted that a few dozen "hotspots" of potential water crises would emerge across the western United States by 2025—many in small to medium-sized inland communities.[60] These smaller communities have limited technical and financial resources to implement DPR. Lacking the economies of scale of larger systems, O&M costs are also relatively higher in small systems. Likewise, monitoring, sampling, and analysis can be challenging and expensive. A further challenge is recruiting, retaining, and paying for qualified expert operators.[61]

Cloudcroft was fortunate in successfully obtaining grants to fund most of its capital costs, including the upgrade to the MBR system at its existing wastewater-treatment plant. Many small communities have elementary or outdated conventional wastewater-treatment facilities that would require significant upgrading just to get started on DPR. Nonetheless, through a combination of fate and willingness to experiment, Cloudcroft, along with Big Spring and Wichita Falls, has paved the way for increased awareness and guidance for DPR as a potential option for smaller communities.

In summary, while DPR has long been discussed in concept, Windhoek remained the world's only operational DPR facility for more than four decades. The first DPR facility in the United States came online in Big Spring, Texas, in 2013. Direct potable reuse took a major step forward in 2015, with the publication of *Framework for Direct Potable Reuse* by a panel of potable reuse experts convened by the WateReuse Research Foundation and three other major water organizations.[62] Arizona, California, Colorado, Florida, New Mexico, and Texas are currently exploring guidance and regulations for DPR. Utilities in several of our case studies—Scottsdale, Denver, Gwinnett County, San Francisco, and Hillsborough County, Florida—are also exploring DPR.

California is the leader in pursuing the scientific basis for DPR.[63] In March 2021, California set a high bar with its draft criteria for DPR. Among other proposed requirements, the sum of the log reductions must be validated for at least a 20-log reduction for enteric viruses, 14-log for *Giardia* cysts, and 15-log for *Cryptosporidium* oocysts. The treatment train must consist of at least four separate treatment processes each for enteric viruses, *Giardia*, and *Cryptosporidium*, although a single treatment process may provide log-reduction credits for more than one pathogen.[64]

With DPR, advanced-treated wastewater can be combined with the raw water supply as input to a drinking-water treatment plant (commonly referred to as *raw-water augmentation*), or introduced directly into the distribution system, bypassing the drinking-water treatment plant (*treated-drinking-water augmentation*). The former option is usually cited as the preferred alternative at this stage of development of the technology and is the option taken at Cloudcroft, Big Spring, and Wichita Falls. Conversely, Windhoek and El Paso have turned to treated-drinking-water augmentation. California is developing regulations for raw-water augmentation scheduled for 2023, with treated-drinking-water augmentation to follow.

Indirect potable reuse is a mature technology compared to direct potable reuse, but DPR appears to have a promising future. With less piping, pumping, and treatment, direct potable reuse has the potential to be more cost effective than indirect potable reuse and has a smaller carbon footprint. In some places, DPR is the only option for potable reuse.

Chapter Thirteen

Small Can Be Beautiful

The future ain't what it used to be.

—Yogi Berra[1]

When Gillette Stadium was being constructed for the New England Patriots in Foxborough, Massachusetts, it became clear that the town would not be able to meet the new stadium's water demands. During games, the number of fans would far exceed the population of this small town twenty-eight miles southwest of Boston. Likewise, Foxborough's wastewater-treatment plant would be swamped by wastewater from the stadium, especially during the peak "halftime flush." A water-reuse system resolved both problems. Wastewater flows generated by the beer-drinking and fast-food-eating fans are captured and run through a membrane bioreactor treatment plant. The recycled water is used for flushing toilets, cooling, and other nonpotable uses at the stadium and an adjacent retail complex known as Patriot Place.[2]

The Minnesota Twins Target Field captures about two million gallons per year of rainfall otherwise drained into the Mississippi River. The rainwater is treated with a filtration system including ultrafiltration, disinfected with chlorine and ultraviolet radiation, and stored in a holding tank. The water is used for field irrigation and wash-down of the lower grandstands, then sent to a municipal wastewater-treatment plant before discharge to the Mississippi River. The system reduces water consumption from the local drinking-water plant and has a water-quality benefit in that untreated stormwater previously flowing to the Mississippi River is now treated. Filtration systems were also installed in fountains throughout the stadium, and fans are encouraged to bring reusable water bottles.[3]

Denver Water has partnered with others to explore the use of recycled bath and shower water for toilet flushing in a pilot project involving forty new homes in a development.[4] Self-contained residential water-treatment systems, roughly the size of a small stacked washer and dryer set, capture water from showers and baths and treat it to remove hair, solid particles, microorganisms, and soaps. The water is then stored until the toilet is flushed. Excess graywater is sent directly into the sewer line. If there's too little water for toilet flushing, the system taps into water lines to make up the difference. Two showers per day will typically meet the toilet-flushing needs for a family of four. The potential water savings are considerable, as much as 25 percent of indoor water use. Denver Water has also built reuse into its new campus, with onsite treatment of wastewater for toilet flushing and landscape irrigation.

At 1,070 feet, Salesforce Tower is the tallest building in San Francisco and the first thing travelers see when approaching the city from any direction, often rising above the city's fog. The building is notable on the inside as well, featuring the largest blackwater-recycling system in a commercial high-rise building in the United States. Wastewater from sources such as rooftop rainwater collection, showers, sinks, toilets, and urinals is collected, treated in an onsite treatment plant in the basement, and recirculated to meet the building's demands for toilet flushing, irrigation, and cooling.[5]

The Blue Hole Primary School in Wimberley, Texas, cautiously opened its doors for classes in the midst of the COVID-19 pandemic in August 2020, with a broad array of design features that reduce its water footprint. Located in the Hill Country near Austin, the school is named for a popular spring-fed swimming spot that the water savings help preserve. Reducing water usage starts with water-saving fixtures, but the school goes well beyond this by using the "One Water" concept that all water is a resource. Captured rainwater and air-conditioning condensate are used to flush toilets and provide irrigation for landscape and school gardens. The school's wastewater is treated and used to irrigate athletic fields through a subsurface drip irrigation system. Bioswales, rain gardens, and walkways with permeable pavers slow down runoff, recharge groundwater, and reduce nonpoint source pollution. All told, the Blue Hole Primary School's unique campus reduces its groundwater usage from the overstressed Trinity Aquifer by 90 percent compared to traditional construction standards. The One Water system not only saves water but also saves money. Lower water and wastewater bills will add up to as much as $1 million (present day) in savings over the course of thirty years. Perhaps most importantly, this water-smart school educates students and visitors about the importance of taking care of their watershed and natural resources. Clear pipes and signage are installed into the architecture of the school to help create an immersive, educational experience.[6]

These are just a few examples of the growing popularity of onsite or decentralized nonpotable water systems. In these systems, local sources of water (roof rainwater, stormwater runoff, air conditioning concentrate, and various types of wastewater) are collected, treated, and used for nonpotable applications at the household, neighborhood/ multi-residential, or commercial scale. Onsite nonpotable reuse avoids the hefty price tag of centralized purple pipe systems and associated disruption to install pipelines in densely developed urban areas.

There are several other drivers for an onsite water system. The most obvious is that it reduces the need for potable water. For example, why flush toilets with treated drinking water, when you can reuse your bath water for this purpose? Green enthusiasts view onsite water systems as part of the movement toward buildings with net-zero water and energy use. Companies see them as a way to achieve coveted green building certifications, such as LEED (Leadership in Energy and Environmental Design), the most widely used green building rating system in the world.

Onsite treated wastewater (along with roof rainwater and stormwater runoff) can be used for a variety of nonpotable uses. The most common indoor use is for toilet and urinal flushing, which can represent approximately 25 percent of the total water demand in a residential building and up to 75 percent of the total water demand in a commercial building. Other potential nonpotable water uses include irrigation, cooling/heating applications, process water, and clothes washers. These additional applications can increase the nonpotable water demand up to

50 percent for residential buildings and up to 95 percent for commercial buildings.[7]

Onsite water-reuse systems distinguish between graywater and blackwater. Graywater includes wastewater from bathroom sinks, showers, bathtubs, clothes washers, and laundry sinks. Blackwater includes water from toilets and urinals, as well as kitchen wastes from dishwashers and kitchen sinks. Kitchen wastes are classified as blackwater because of the associated organic matter, oil and grease, and uncooked food with bacteria on it. Nonlaundry utility sinks are also usually considered sources of blackwater.

Graywater reuse offers the potential for substantial savings of potable water for residential and multi-residential applications. Many commercial facilities, however, do not generate enough graywater to justify use for toilet flushing or irrigation. Exceptions are hotels, gyms, and other commercial facilities with showers or laundry onsite.[8]

Graywater is cleaner than regular wastewater but still may contain traces of dirt, grease, household chemicals used for personal hygiene and house cleaning, and any medications and waste products disposed of in sinks. Some level of treatment is needed for most uses. Disinfection of graywater is needed for uses that may lead to human exposure to pathogens that become airborne or are ingested with small amounts of water.

At the household scale, simple graywater systems that require little energy and maintenance can be used to water lawns and flowerbeds using subsurface irrigation. Studies of the effects of graywater use on soil and plant health have found that selected plants (e.g., avocado trees) are sensitive but that there are no major concerns overall.[9] It's advisable for homeowners with these systems to use biodegradable soaps and "plant-friendly" products without a lot of salt, boron, or chlorine bleach.[10] If water conservation is the primary objective for graywater irrigation, then it makes sense to first consider ways to reduce outdoor water use.

Toilet flushing is a popular use of reclaimed graywater in multi-residential buildings or neighborhoods. Flushing toilets can account for a significant percentage of water use and is relatively constant throughout the year. Graywater systems for toilet flushing require dual plumbing with a connection to potable water and backflow preventers that require annual inspection. Graywater used for toilet flushing clearly doesn't need to be drinking-water quality but should be disinfected. Aesthetic concerns about off-color water also may need to be addressed.[11]

Concerns about pathogens in graywater used for flushing toilets might come as a surprise given that's how we get rid of our excrement. Scientists have found that incidental exposure can arise from aerosols forced upward by the turbulence after flushing. Experts call it the "toilet plume." For all the attention paid to the surface of public toilet seats—as we carefully lay down thin covers in stalls—the risk of germ transmission from skin contact may be relatively small compared with what happens when you flush. The risks of graywater use are likely very low, but disinfection is a precautionary measure, particularly given the frequency of exposure.[12]

Regulations of graywater differ by state and may even differ within a state between its plumbing codes and other regulations.[13] Requirements also depend on the application. At the single-residence scale, family members or housemates are already exposed to each other in many ways, so use of graywater adds minimal additional risk. Local agencies usually simply identify best management practices for operation and maintenance. At the neighborhood/multi-residential scale, maintenance is often managed by contractors or on-site staff. Depending on state or local regulations, a licensed operator may be required for some wastewater-reuse systems but is generally not required for graywater systems.[14] Some requirements for indoor applications require daily monitoring of indicator organisms (e.g., total coliforms or *E. coli*). An alternative involves continuous monitoring of parameters, such as turbidity, that serve as surrogates for pathogens.[15]

The value of onsite reuse depends on the situation, as the title of this chapter suggests, in a take-off from the popular book *Small Is Beautiful* by E. F. Schumacher. For example, onsite reuse for irrigation may either save water or increase water use, depending on whether the water is applied to an existing or new irrigation use. The worst-case scenario is that using graywater encourages landscaping that is inappropriate for local climate conditions and not sustainable in the long term. The value of onsite reuse also depends on the extent that wastewater effluents from the wastewater-treatment plant currently serve beneficial uses. If the effluent is critical to downstream water supply or supports vital aquatic habitat in stressed rivers, then onsite reuse may essentially be "robbing Peter to pay Paul."

If onsite reuse becomes widely adopted in an area, potential unintended consequences need to be considered in overall urban water-systems planning. Use of onsite graywater reduces flows and increases pollutant concentrations in the wastewater. With less water to push waste through the sewers, sewage can become stagnant, causing corrosion and stinky neighborhoods. Onsite reuse systems are unlikely to have positive returns on investment in the short term but can be wise long-term investments. Concerns about private control of onsite systems and their long-term management also need to be addressed.[16]

With limited freshwater resources, San Francisco has taken up the mantle of onsite water reuse in a major way with plenty of incentives. The city is surrounded on three sides by saltwater. The climate is semi-arid. Fog, not rainfall, is the area's most distinctive characteristic during the long, dry summers. At the same time, all of the city's unused treated wastewater goes to the ocean or bay, without benefitting any downstream users or environmental needs. The city also has more wastewater than it can handle during storms, when its combined sewer systems overflow from the combination of rainwater and sewage.

Local streams, springs, and wells sufficed to meet the city's needs until the 1848 gold rush led to a booming metropolis almost overnight. San Francisco became an "instant city with an instant water problem," comments historian Norris Hundley.[17]

Today, San Francisco gets most of its water from Hetch Hetchy Reservoir. Located 167 miles away in Yosemite National Park, the reservoir is filled by snowmelt that feeds the Tuolumne River. San Francisco's water rights date back to 1901, when city leaders were desperately searching for a new supply. As the story goes, the claim was tacked to an oak tree along the riverbank. Wilderness advocates, led by John Muir, fought the damming of the magnificent glacier-sculpted valley, but the battle was eventually lost. In 1913, President Woodrow Wilson signed the Raker Act, granting San Francisco rights to build a dam and flood the valley.[18]

The water from Hetch Hetchy flows through a chain of smaller reservoirs and pipelines that carry it directly to San Francisco. The drinking water provided is among the purest in the world, and the system is notable for being almost entirely gravity fed, requiring almost no fossil-fuel consumption to move water from the mountains to the tap.[19]

Frequent and severe droughts during the past two decades, as well as climate uncertainties, have heightened concerns about San Francisco's dependence on Hetch Hetchy Reservoir, which is also vulnerable to earthquakes and wildfires. The transported water crosses three major earthquake faults—the Calaveras, Hayward, and San Andreas. Environmentalists also press for releasing more water downstream from Hetch Hetchy for struggling salmon and steelhead populations and to increase freshwater flows into the Bay Delta. These concerns have incentivized the city to encourage conservation and diversify its water portfolio.

Part of San Francisco's diversification makes strategic use of its limited groundwater through the Regional Groundwater Storage & Recovery project, scheduled for completion in 2021. The project uses a groundwater basin south of San Francisco as a savings account. Pumping is limited to dry years when the groundwater is needed most. During normal and wet years, other water sources (mostly Hetch Hetchy) are used in lieu of groundwater pumping, allowing aquifer storage to build up during this time.[20]

In the process of diversifying its water sources, San Francisco has become the undisputed national leader in onsite water reuse. This leadership role follows an early venture several decades ago when recycled wastewater was used at the city's most famous park in the nation's first urban water-reuse system.

Visitors to Golden Gate Park with its bucolic lakes, tree groves, and gardens would likely be surprised to learn that the area was once covered by windswept sand dunes. They would be even more surprised to learn that raw sewage supplemented groundwater to irrigate the park in the early 1900s. The sewage was transported through open ditches and stored in septic tanks. As the city grew around the park, people increasingly complained about the noxious sewage odors.[21]

The city abandoned the use of raw sewage in 1932 upon completion of a nearby activated-sludge treatment plant. The chlorinated plant effluent was used to create a chain of lakes connected by artificial brooks and waterfalls. The treatment plant eventually went out of compliance as stricter regulations were enacted in the 1970s, and it was shuttered in 1982. At that point, wastewater recycling in San Francisco went dormant, and the park turned to San Francisco's precious drinking-water supplies (mostly from groundwater) for irrigation and to fill its lakes.[22]

The San Francisco Public Utilities Commission (SFPUC) provides water, wastewater, and electric-power services to almost 3 million customers in San Francisco and three other Bay Area counties (Alameda, Santa Clara, and San Mateo). In 2017, as part of its effort to rely more on local water sources, the SFPUC broke ground for a modern advanced-wastewater treatment plant using membrane filtration, reverse osmosis, and UV disinfection. The advanced-treatment plant, which is connected to an existing secondary-treatment plant, is scheduled to be completed in 2022, marking the return of recycled water to Golden Gate Park after an absence of nearly four decades. Recycled water will also be used for irrigating golf courses and other landscaped areas on the city's west side. The level of treatment is more rigorous than usual for nonpotable uses in order to reduce the high salinity and ammonia to acceptable levels for use in parks and to fill lakes. The advanced-treatment system also may eventually be a step toward future potable reuse.[23]

The nation's most cutting-edge onsite water-reuse program began when the SFPUC was designing their new headquarters building in downtown San Francisco. Billing itself as providing high-quality, efficient, and reliable water, power, and sewer services in a sustainable manner, the utility wanted to walk the talk with their building design. One way to do this was to install an onsite system to reclaim and treat the building's wastewater for use with toilets and urinals.

Beginning with its opening in 2012, all of the building's wastewater is treated using a constructed wetland system (after primary treatment). Referred to as the Living Machine, this is essentially a series of large planter boxes located in the sidewalks surrounding the headquarters, as well as in the lobby. The system treats approximately five thousand gallons of wastewater per day and then distributes the treated water for toilet and urinal flushing. The system blends both function and aesthetics in reducing the building's total water use. The building also captures rainwater from the roof and from the outdoor play area of the children's daycare center. This water is treated and used for subsurface irrigation for plantings and street trees.[24]

In 2012, San Francisco became the nation's first city to adopt an ordinance for the installation and operation of private onsite nonpotable systems.[25] In 2015, it became mandatory for new developments with over

250,000 square feet of floor area to install and operate an onsite water system for toilet/urinal flushing and irrigation using available graywater, rainwater, and foundation drainage. Recent proposals would require future large commercial buildings to treat and reuse blackwater from toilets and kitchen sinks.[26] The SFPUC also provides grant funding for installing onsite water systems on a voluntary basis, or for exceeding the minimum requirements. Breweries are also eligible for grant funding to collect, treat, and reuse brewery process water for applications such as tank and bottle rinses.

At a smaller scale, the city encourages simple laundry-to-landscape systems. California allows household laundry-to-landscape systems without a permit as long as they follow specific design guidelines. A valve makes it easy to divert laundry water from irrigation to the sewer system—for example, when running a load with bleach or diapers. To sweeten the deal, the city offers residents a $125 subsidy toward the $175 cost of the installation kit, a free in-home consultation with a graywater expert, a free workshop on how to safely and properly install the kit, and access to a free toolkit for do-it-yourself installation.[27]

Laundry-to-landscape systems are a feel-good part of more sustainable living, but their effectiveness at reducing water use depends on whether the graywater is used for previously irrigated or expanded irrigated landscape. There's also the problem of feeling like it's okay to use more water for other purposes (take longer showers), in light of all the water savings coming from reuse of laundry water for irrigation. Of course, if you haven't already done so, it pays to replace that old water-hogging clothes washer with a modern one before moving to laundry-to-landscape. A national study of residential water uses in 2016 found that high-efficiency clothes washers accounted for the largest per-capita reduction in indoor water use, followed by toilets.[28]

The SFPUC is also experimenting with potable reuse. Lacking a sufficient surface-water body or aquifer to serve as an environmental buffer, San Francisco's only option is direct potable reuse. The SFPUC is exploring this option using its headquarters as a testing ground. Some of the recycled water produced by the constructed wetland treatment system is further treated using ultrafiltration, RO, and UV radiation. This purified water is tested for quality and then returned to the nonpotable water system.[29]

San Francisco's water-reuse program is impressive. But what is most remarkable is how the utility has spearheaded the development of national guidance for onsite nonpotable systems under the leadership of Paula Kehoe, its director of water resources. When the SFPUC undertook its onsite water-reuse program, guidance on treatment requirements, permitted uses, and other factors was virtually nonexistent. Such guidance is critical to foster adoption of onsite water systems, not just in San Francisco but also across the country.

In May 2014, the SFPUC brought representatives of utilities and public health agencies from across North America to San Francisco to discuss onsite water systems. The meeting led to a step-by-step guide for developing a local program to manage onsite water systems.[30] This was a good start. A follow-up multiyear study tackled perhaps the biggest challenge facing implementation of onsite water systems—the lack of risk-based water-quality guidelines.[31] The SFPUC also sponsored a bill directing the California Water Board to establish uniform statewide risk-based criteria for each type of onsite recycled-water use by 2022.

On March 22, 2016, at a White House Water Summit to honor World Water Day, the SFPUC and US Water Alliance announced a commitment to accelerate the development of onsite water-reuse projects across the country by creating the National Blue Ribbon Commission for Onsite Nonpotable Water Systems. Comprised of about thirty members and chaired by Kehoe, the commission develops guidance and model policies and identifies research needs.[32]

Other cities have taken up the mantle of onsite water reuse. Santa Monica, California, waives building-permit fees. New York City provides wastewater allowances to qualified properties with onsite water systems.[33] Austin, Texas, is perhaps the most ambitious. In 2018, this fast-growing city adopted a one-hundred-year "Water Forward Plan" to address future water supply in the face of profound impacts of climate change and a growing population. It's an adaptive plan to be updated on a five-year cycle. Water reuse is a major component, driven in part to avoid the common tactic of the city reaching out its tentacles to take water from other places. Austin Water (the water and wastewater utility) anticipates that community-scale onsite water reuse will account for one-third of all new water supplies by fostering reuse of all water flows within the city. This includes air conditioner condensate, rainwater, stormwater, and blackwater. This ambitious undertaking plans to scale up quickly, with a goal to capture and treat twenty times more water from buildings than any other city in the United States by 2040.[34]

Onsite nonpotable reuse is part of broader possibilities for future use of decentralized systems. In the book *Water 4.0*, David Sedlak notes that centralization has been the "big idea" behind urban water systems since the pioneering Roman aqueducts and sewer systems. Decentralized systems, he says, represent a significant opportunity to transform the way water is managed, offering the advantages of increasing flexibility, reducing energy consumption, and lowering the costs of replacing, expanding, and upgrading water and sewage infrastructure.[35]

Conventional wastewater-treatment plants generally are located downgradient in urban areas, permitting gravity wastewater flow to the treatment plant, but the demand for reclaimed water generally lies upgradient. Locating smaller distributed (satellite) treatment plants closer to areas of demand could substantially reduce the energy costs for pumping, as well as delay or mitigate the need for expensive infrastructure expansion. Alternately, scalping or sewer-mining plants can treat raw sewage by tapping into existing regional sewer lines, producing recycled water for local use before sending the residuals back into the sewer system.

If wisely chosen and cooperatively planned, decentralized systems (onsite systems and satellite plants) can help address the cost of fixing our water, wastewater, and urban drainage systems. The overall costs are staggering. More than $200 billion may be needed over the next two decades to maintain and upgrade publicly owned wastewater pipes and plants in the United States.[36]

Decentralized wastewater systems, including sensors and autonomous control systems, are an active area of research.[37] Among the focus areas are membrane bioreactors, which combine a low-pressure membrane process, such as microfiltration or ultrafiltration, with a biological process. These systems avoid the need for large clarifiers (settling tanks) to remove suspended solids. Their compact size and ease of automation make them useful for decentralized systems.[38]

An emerging technology known as a Staged Anaerobic Fluidized Membrane Bioreactor (SAF-MBR) illustrates the opportunities. Like conventional wastewater-treatment plants, MBRs typically rely on *aerobic* (oxygen-using) processes, which consume large amounts of energy. Stanford University has been working with South Korean researchers on an alternative using *anaerobic* processes.[39] Water is treated in two stages as it moves through the SAF-MBR. In the first stage, granular activated carbon is suspended in the reactor by the upward velocity of the wastewater being treated. Bacteria growing on the activated carbon anaerobically treats the wastewater. In the second stage, hollow fiber membranes separate the remaining solids (and most pathogens) from the water as it leaves the system. The solids remaining inside begin to hydrolyze, making them more accessible to bacteria. The SAF-MBR system generates energy from methane, produces fewer biosolids than aerobic processes, and appears to have superior capability at reducing contaminants of emerging concern.[40]

In summary, small(er) systems at various scales can be "beautiful" when thoughtfully considered and integrated into water and wastewater systems. As in all types of water reuse, it's not a silver bullet, but it is a potential way to strengthen future water supplies. The possibilities are continuing to take shape.

Chapter Fourteen

One Water

By employing sophisticated public communication strategies and state-of-the art treatment technologies, the forces behind potable water recycling now appear to be unstoppable.

—David Sedlak, *Water 4.0*[1]

In 2013, the Water Environment Federation formally began using the term *water resource recovery facility*, in place of *wastewater-treatment plant*.[2] The name change signifies how wastewater-treatment plants are evolving from simply a means of protecting surface waters from pollution to systems that recover water, energy, and nutrients from the sewage, while also protecting the environment. Wastewater is increasingly viewed as a resource to be used rather than as a waste to be disposed of. Making cities water wise is part of growing attention to building cities that are more resilient, livable, and sustainable.

As a resource, wastewater is increasingly viewed through a "One Water" lens. Historically, water, wastewater, and stormwater have been compartmentalized as distinct and isolated management areas, with little or no collaboration among the separate agencies. This fragmented, siloed approach impeded efforts to identify and implement innovative opportunities to enhance the reliable supply of clean, sustainably sourced freshwater. The One Water movement recognizes the vast potential for collaborative management when all aspects of water are viewed as a single integrated system. Instead of compartmentalizing water by source—groundwater and surface water, stormwater, and wastewater—consider it all as "water."

The hallmarks of One Water are the following:

- a perspective that all water has value;
- a focus on achieving multiple benefits;
- approaching decisions with a systems mindset;
- utilizing watershed scale thinking;
- finding rightsized solutions; and
- relying on partnerships and inclusion.[3]

A central tenet of the One Water movement is to encourage treating water to a level that is fit-for-purpose—in other words, matching the availability and quality of water with the requirements for its intended use. One Water advocates emphasize that the water present today is the same water that existed when the dinosaurs roamed the Earth and will be the same water that exists for future generations.

Water reuse fits neatly into the One Water movement, as well as being a natural evolution of how water utilities can meet demands in the face of a growing population and climate change. By offering renewable and more drought-proof supplies, it enhances overall dependability and bolsters independence. Wastewater is an appealing resource because utilities don't have to go out and buy it—they already own it.

Monterey County, California, is a good example of One Water in action. In 2017, the regional wastewater agency was renamed Monterey One Water. Two decades earlier, the agency had pioneered the use of recycled wastewater for irrigating food crops in the Salinas Valley (see chapter 2). The name change reflected the use of three additional water sources for recycling—urban stormwater runoff, drainage water from agricultural fields, and food industry processing water from washing ready-to-eat salad mixes and vegetables. These new sources not only provide additional water for reuse but also reduce wastewater discharges into the sensitive Monterey Bay National Marine Sanctuary.[4]

Monterey One Water also added advanced water treatment to address the overreliance of area residents on the Carmel River and Seaside Groundwater Basin for public supply. Monterey One Water partnered with other utilities to form Pure Water Monterey to replenish the Seaside Groundwater Basin with advanced-treated water for indirect potable reuse and allowing reduced Carmel River diversions. Depending on the need, Monterey One Water determines how much of each of its four water-recycling sources will receive advanced water treatment for potable reuse, and how much will get tertiary treatment for agricultural irrigation in the Salinas Valley.[5]

Wastewater reuse has lots of growing room. Nationwide, approximately thirty-three billion gallons of wastewater are treated each day. Most is returned to the environment as treated effluent, while an estimated 2.2 billion gallons per day (6.6 percent) is recovered for reuse.[6] While much of the wastewater discharged to streams and rivers is serving beneficial environmental purposes and downstream users, fully one-third of it is directly discharged to the ocean—an amount equivalent to about 30 percent of the U.S. public water supply.[7]

Reusing municipal wastewater has possibilities that go beyond the traditional applications. For example, the City of Santa Rosa, California, pumps tertiary-treated wastewater forty miles uphill, where it's injected underground at the Geysers geothermal field to generate electricity for one hundred thousand households.[8] Also of growing interest is reusing various industrial and agricultural wastewaters, as well as produced waters from oil and gas operations.

Water utilities are inherently conservative and risk adverse. They tend to favor incremental innovation to assure that above all else, high-quality, low-cost water is consistently delivered to their customers.[9] Public concerns about the risks and costs of treated wastewater reinforce this conservatism. Nonetheless, early adopters of reuse have laid substantial groundwork for utilities who prefer a "me second" approach.

Water-reuse technologies have the potential to help address the massive investments needed for repairs, replacements, and upgrades of aging infrastructure.[10] Among the challenges facing water reuse in meeting this need are the regulatory environment, financial stresses in many communities, and tradeoffs with water conservation.

Regulatory Environment: Before investing in potable reuse, water utilities seek to minimize regulatory uncertainty. State regulatory agencies together with national nonprofit associations—such as the WateReuse Association (and its state chapters), American Water Works Association, National Water Research Institute, and the Water Research Foundation—have worked for decades developing best practices, standards, laws, and manuals, as well as advancing public acceptance of recycled water. Various scientific enterprises also help meet the challenges. Since 1999, the Urban Water Center at Colorado State University has been a leader in graywater reuse. In 2011, Stanford, UC Berkeley, Colorado School of Mines, and New Mexico State University launched Re-Inventing the Nation's Urban Water Infrastructure (ReNUWIt), a major collaborative effort funded by the National Science Foundation.

Supreme Court Justice Louis Brandeis famously called states the "laboratories of democracy" to test new ideas before they are universally adopted. The concept applies well to potable reuse. Given the ongoing research and technological innovation underway in water reuse, "there may be a real danger of locking in, by way of national regulation, what may become obsolete technology," cautions Tracy Mehan, executive director of government affairs for the American Water Works Association.[11]

Mehan's view appears to be widely shared. The EPA has studied potable reuse since the 1970s and published guidelines for potable reuse at various times over the years. Water-recycling standards have remained the responsibility of the states, but must meet EPA drinking-water standards. In 2020, the EPA released its National Water Reuse Action Plan as a "call to action" for greater reuse.[12] The action plan is all about collaboration among federal, state, tribal, local, and water-sector partners. For the foreseeable future, states and associations will likely remain the leaders and innovators in water reuse with the EPA serving largely as a facilitator and resource hub, along with its traditional roles under the Safe Drinking Water and Clean Water Acts.

Strengthening some existing federal laws, however, could help address weaknesses. A key area is controlling hazardous substances in the environment through the Toxic Substances Control Act. When enacted in 1976, this largely toothless act grandfathered in chemicals already on the market and made it very difficult to regulate new chemicals. As a result, wastewater agencies have to deal with toxic chemicals that should have been restricted. Today's widespread contamination by PFAS is an extraordinary example of the failure of the act. The Toxic Substances Control Act was reformed in 2016, but the effectiveness of these changes remains to be seen.[13] Monitoring under the Clean Water

Act's National Pretreatment Program also needs updating for better source control of chemicals of concern to potable reuse.

Financial Challenges: Utilities have to balance three competing financial concerns: maintaining affordable rates for their customers, investing in infrastructure, and ensuring their own fiscal stability.[14] Areas with shrinking populations and/or declining incomes amplify these challenges. A smaller revenue base makes it difficult to maintain the fixed costs of day-to-day operations, let alone finance new infrastructure. In particular, many small water systems and those serving disadvantaged communities are perennially on shaky financial footing. COVID-19 compounded these challenges, as water utilities faced lower revenues and numerous unpaid and late water bills. It has been estimated that low-income households spend an average of 12 percent of their disposable income, and work about ten hours at minimum wage, to pay for monthly water and sewer services.[15]

Water reuse as part of more decentralized systems located at or near the point of use has the potential to be less expensive than upgrades to large, centralized systems, as well as provide greater financial and operational flexibility. These possibilities present an opportunity to address the needs of vulnerable and marginalized communities. Investment through federal and other programs for disadvantaged communities could help cover some costs. Ideally, these facilities could provide jobs and support local economic development, as well as create greener public spaces. But there are also challenges in overcoming perceptions that poorer communities are getting second-rate water through reuse, compounded by entrenched concerns by various ethnic communities about the overall safety of drinking water.

Water Conservation: Water conservation is the cheapest and most energy efficient source of "new" water supply. In recent decades, conservation programs have been very effective in reducing overall water demand. Among the success stories are the EPA WaterSense Program, which promotes and certifies water-efficient products, "Cash for Grass" and other incentives for cutting outdoor water use, and rebates by municipalities for replacing indoor plumbing fixtures with more efficient models. And there's still plenty of room for additional conservation. For example, less than half of American homes have installed highly efficient toilets or clothes washers, two of the greatest uses of water indoors.[16]

Water reuse and conservation can be complementary but also require special consideration. Conservation affects the future demand for potable reuse, as well as the amount of wastewater available for reuse. Water-use projections have consistently underestimated the effectiveness of conservation and overestimated future water demands.[17] Mandated or voluntary indoor water conservation during droughts reduces flows to wastewater-treatment plants, potentially leading to less recycled water than anticipated. As a result, treated wastewater may be less drought resistant than planned.[18]

Gilbert Trejo, chief technical officer of El Paso Water and 2021 president of the WateReuse Association, describes potable reuse as a long journey. His advice to the many utilities across the country that may ultimately benefit from the practice is to start small with perhaps a purple pipe project, gradually and regularly educate the public about water reuse, and continue to build water-reuse capability with time. But get started sooner rather than later, and play the long game. While the needs may seem far off in the future, even a decade is "imminent" in the water-planning realm.[19]

Potable reuse appears to have a bright future. Southern California cities have widely adopted it to reduce dependence on more expensive and less reliable imported water. Other water-stressed communities in the desert southwest as well as cities along the East Coast have also turned to potable reuse, the latter to address water quality as well as water availability. Expansion of potable reuse to other areas seems inevitable, with indirect potable reuse a relatively mature technology and direct potable reuse the next frontier.

The water reuse community has come a long way in honing its messages. Adherents recognize that gaining support for potable reuse is not simply a matter of experts choosing what they perceive as the most desirable solution and then seeking ways to convince the public. Elected officials, opinion leaders, as well as the general public need to be involved early, and repeatedly, in a meaningful and open way. While water reuse plays out in different ways, depending on the local situation, we've found one constant: It's hard to find a more dedicated and enthusiastic group than those involved in the water-reuse world. People feel they're part of something important and are more than happy to talk about it with anyone who is interested.

Notes

CHAPTER ONE

1. Quoted in R. Glennon, *Unquenchable: America's Water Crisis and What to Do about It* (Washington, DC: Island Press, 2009), 205.

2. R. Rivard, "A Brief History of Pure Water's Pure Drama," *Voice of San Diego*, September 17, 2019, www.voiceofsandiego.org/topics/government/a -brief-history-of-pure-waters-pure-drama.

3. J. Hill, "Dry Rivers, Dammed Rivers and Floods," *Journal of San Diego History* 48, no. 1 (2002).

4. D. Walker, *Thirst for Independence: The San Diego Water Story* (San Diego: Sunbelt Publications, 2004), 14–16.

5. Ibid., 66.

6. Ibid., 65.

7. Ibid., 70.

8. R. Stayton, "Sludge Busters," *Popular Science* (February 1987): 43–44.

9. "San Diego—the Only City in the World to Mix Drinking Water with Sewage," *San Diego Reader*, May 8, 1997, www.sandiegoreader.com/ news/1997/may/08/cover-san-diego-only-city-mix-drinking-water-sewag.

10. Stayton, "Sludge Busters."

11. "San Diego—the Only City."

12. Western Consortium for Public Health, *The City of San Diego Total Resource Recovery Project: Health Effects Study—Final Summary Report* (Oakland: Western Consortium for Public Health, 1992).

13. National Research Council, *Water Reuse: Potential for Expanding the Nation's Water Supply Through Reuse of Municipal Wastewater* (Washington, DC: National Academies Press, 2012), 75.

14. National Research Council, *Issues in Potable Reuse: The Viability of Augmenting Drinking Water Supplies with Reclaimed Water* (Washington, DC: National Academy Press, 1998), 170.

15. Patricia Tennyson, interview, November 4, 2019.

16. J. Leovy, "Reclaimed Waste Water May Ease State's Thirst," *Los Angeles Times*, August 17, 1997, www.latimes.com/archives/la-xpm-1997-aug-17-mn-23514-story.html.

17. A. Espinola, "How One Utility Won Public Support for Potable Reuse," *AWWA*, January 21, 2016, www.awwa.org/AWWA-Articles/how-one-utility-won-public-support-for-potable-reuse.

18. S. M. Siegel, *Troubled Water: What's Wrong with What We Drink* (New York: Thomas Dunne, 2019), 217.

19. S. M. Katz and P. Tennyson, "Coming Full Circle: Craft Brewers Demonstrate Potable Reuse Acceptance," *Journal AWWA* 110, no. 1 (2018): 62–67.

20. K. Balint and G. Braun, "Public Distaste Stalls 'Toilet-to-Tap,'" *San Diego Union-Tribune*, December 9, 1998, A1, A8.

21. Katz and Tennyson, "Coming Full Circle."

22. M. Quartiano, *Toilet to Tap: San Diego's "Pipe Dream"* (self-published manuscript, 2006), 26–27, available at San Diego Public Library.

23. Balint and Braun, "Public Distaste."

24. K. Balint, "Water from (Gulp!) Where?" *San Diego Union-Tribune*, July 6, 1997, B1, B6.

25. Katz and Tennyson, "Coming Full Circle."

26. "San Diego—the Only City."

27. S. Jobling, M. Nolan, C. R. Tyler, G. Brighty, and J. P. Sumpter, "Widespread Sexual Disruption in Wild Fish," *Environmental Science & Technology* 32, no. 17 (1998): 2498–2506.

28. U.S. Environmental Protection Agency, *Mainstreaming Potable Water Reuse in the United States: Strategies for Leveling the Playing Field* (ReNUWIt and the Johnson Foundation, 2018), 10, www.epa.gov/sites/production/files/2018-04/documents/mainstreaming_potable_water_reuse_april_2018_final_for_web.pdf.

29. "Muriel Watson Obituary Guest Book," *San Diego Union-Tribune*, May 4, 2012.

30. M. Potter, "Good Old Boys and Girls," *San Diego Reader*, September 10, 1998, https://perma.cc/G38Y-JN73.

31. National Research Council, *Issues in Potable Reuse*, 3, 15.

32. San Diego County Grand Jury 1998/1999, *Water for the City of San Diego* (1999), 5.

33. Balint and Braun, "Public Distaste."

34. San Diego City Council, *Resolution Number R-291210*, adopted January 19, 1999, https://docs.sandiego.gov/council_reso_ordinance/rao1999/R-291210.pdf.

35. Balint and Braun, "Public Distaste."

CHAPTER TWO

1. L. A. Stevens, *The Town That Launders Its Water* (New York: Coward, McCann & Geoghegan, 1971): 55.

2. Stevens, *The Town That Launders Its Water*.

3. Ibid., 41–42.

4. Ibid., 58.

5. Ibid., 60.

6. G. Hill, "Purified Sewage Used to Fill Swimming Pool in California," *New York Times*, July 9, 1965, 32.

7. In 1997, the treatment process was upgraded from activated sludge to the Bardenpho biological nutrient removal process.

8. "About Us," Santee Lakes, accessed June 10, 2021, www.santeelakes.com/about-us.

9. City of San Diego, *Water Reuse Study: Water Reuse Goals, Opportunities & Values* (2004), 54, www.sandiego.gov/sites/default/files/legacy/water/pdf/purewater/aa1wp.pdf.

10. T. R. Holliman, "Reclaimed Water Distribution and Storage," in *Wastewater Reclamation and Use*, ed. T. Asano (Lancaster, PA: Technomic, 1998), 395.

11. "Irvine Ranch Water District: An Overview" (February 2020), www.irwd.com/images/pdf/about-us/factsheet.pdf.

12. Keith Lewinger, interview, April 26, 2021; M. Peterson, "Purple Pipe Means Recycled Water. Why Purple?" *KPCC*, March 10, 2014, www.scpr.org/blogs/environment/2014/03/10/16035/purple-pipe-means-recycled-water-why-purple.

13. John Fabris and Mark Tettemer, IRWD, written communication, April 22, 2021.

14. D. G. Metzger, *Geology in Relation to Availability of Water along the South Rim Grand Canyon National Park Arizona* (Washington, DC: U.S. Geological Survey, 1961).

15. C. Graf, "After 90 Years of Reusing Reclaimed Water in Arizona, What's in Store?" *Arizona Water Resource* 24, no. 4 (2016): 6.

16. D. Sedlak, *Water 4.0: The Past, Present, and Future of the World's Most Vital Resource* (New Haven, CT: Yale University Press, 2014), 190, 195.

17. Graf, "After 90 Years."

18. S. Montanari, "Fixing the Grand Canyon's Aging Water Pipeline Won't Be Easy—But It's Necessary," *National Geographic*, May 4, 2021, www .nationalgeographic.com/travel/article/vital-repairs-to-grand-canyon-water -pipeline-will-affect-tourists?mc_cid=e87a91e407&mc_eid=b76a339123.

19. "Upper Occoquan Service Authority," accessed June 10, 2021, www .uosa.org/IndexUOSA.asp.

20. Sedlak, *Water 4.0*, 202–3.

21. J. B. Rose et al., "Reduction of Enteric Microorganisms at the Upper Occoquan Sewage Authority Water Reclamation Plant," *Water Environment Research* 73, no. 6 (2001): 711–20.

22. H. J. Ongerth and J. E. Ongerth, "Health Consequences of Wastewater Reuse," *Annual Reviews in Public Health* 3 (1982): 419–44.

23. S. Harris-Lovett and D. Sedlak, "The History of Water Reuse in California," in *Sustainable Water*, ed. A. Lassiter (Berkeley: University of California Press, 2015), 222.

24. M. F. Jaramillo and I. Restrepo, "Wastewater Reuse in Agriculture: A Review about Its Limitations and Benefits," *Sustainability* 9, no. 10 (2017): 1734.

25. L. E. Lesser et al., "Survey of 218 Organic Contaminants in Groundwater Derived from the World's Largest Untreated Wastewater Irrigation System: Mezquital Valley, Mexico," *Chemosphere* 198 (May 2018): 510–21.

26. A. W. Olivieri et al., *Expert Panel Final Report: Evaluation of the Feasibility of Developing Uniform Water Recycling Criteria for Direct Potable Reuse* (Sacramento, CA: Prepared by the National Water Research Institute for the State Water Resources Control Board, 2016), 16.

27. U.S. Environmental Protection Agency, *Guidelines for Water Reuse* (Washington, DC: 2012), 3–11.

28. Engineering-Science, *Monterey Wastewater Reclamation Study for Agriculture—Final Report* (Monterey, CA: Monterey Regional Water Pollution Control Agency, 1987).

29. "Welcome to the Camden County MUA," Camden County Municipal Utilities Authority, accessed June 10, 2021, www.ccmua.org.

30. A. Farr, "The Mission Beyond the Utility: 2020 WF&M Award Winner Andy Kricun," *Water Finance & Management*, December 14, 2020, https:// waterfm.com/the-mission-beyond-the-utility-andy-kricun.

31. "Community Resources," Camden County Municipal Utilities Authority, accessed June 10, 2021, www.ccmua.org/index.php/resources-for-our -community.

CHAPTER THREE

1. J. F. Kennedy, "Text of President Kennedy's Special Message to Congress on Natural Resources," *New York Times*, February 24, 1961, 12.

2. Some additional recharge occurs through unlined portions of the San Gabriel River. The Rio Hondo is lined with concrete throughout the area and does not recharge the aquifer.

3. T. A. Johnson, "Groundwater Recharge Using Recycled Municipal Waste Water in Los Angeles County and the California Department of Public Health's Draft Regulations on Aquifer Retention Time," *Ground Water* 47, no. 4 (2009): 496–99.

4 . "Fresno-Clovis Regional Wastewater Reclamation Facility (RWRF)," City of Fresno, accessed June 10, 2021, www.fresno.gov/publicutilities/facilities-infrastructure/fresno-clovis-regional-wastewater-reclamation-facility-rwrf.

5. S. Harris-Lovett and D. Sedlak, "The History of Water Reuse in California," in *Sustainable Water*, ed. A. Lassiter (Berkeley: University of California Press, 2015), 223.

6. P. Tennyson and A. Mackie, "Where Did 'Toilet-To-Tap' Come From?" *Clean Water* 5 (2020): 38, www.kelmanonline.com/httpdocs/files/CWEA/cleanwater-issue5-2020/index.html.

7. A. Goldman, "Smaller Water Reclamation Plan Appeases Miller Beer," *Los Angeles Times*, February 9, 1996.

8. B. Hudson, "Mixed Reviews for Water Reclamation Plan: Miller Brewery and Other Opponents of the Project Say It Could Pose Health Risks. Environmentalists and Water Agencies Embrace It as a Way to Help 'Drought-Proof' the San Gabriel Valley," *Los Angeles Times*, December 12, 1993, www.latimes.com/archives/la-xpm-1993-12-12-ga-1066-story.html.

9. A. Sklar, "Toilet-to-Tap," *Los Angeles City Historical Society Newsletter* 48, no. 3 (Fall 2015): 1, 8–10.

10. S. M. Siegel, *Troubled Water: What's Wrong with What We Drink* (New York: Thomas Dunne, 2019), 215.

11. Tennyson and Mackie, "Where Did 'Toilet-to-Tap' Come From?"

12. Hudson, "Mixed Reviews."

13. Sklar, "Toilet-to-Tap."

14. J. Nelson, "'Clown Activist' Jailed, Facing Prison," *San Bernardino Sun*, October 21, 2010.

15. Sklar, "Toilet-to-Tap."

16. F. Clifford, "Storm Brews Over Prospect of Recycled Water in Beer: Sewage: Miller Co. in Irwindale is Suing to Halt $25-Million Project. District Dismisses Concerns," *Los Angeles Times*, September 14, 1994, www.latimes.com/archives/la-xpm-1994-09-14-mn-38525-story.html.

17. A. Goldman, "Smaller Water Reclamation Plan Appeases Miller Beer," *Los Angeles Times*, February 9, 1996.

18. J. L. Sax, "The Public Trust Doctrine in Natural Resource Law: Effective Judicial Intervention," *Michigan Law Review* 68 (1970): 471–566.

19. D. Green, *Managing Water: Avoiding Crisis in California* (Berkeley: University of California Press, 2007), 35.

20. M. Cone, "DWP Agrees to Take Less Mono Lake Water Resources: City Plans to Make Up Loss in Reclamation Effort," *Los Angeles Times*, December 14, 1993, A1.

21. J. Krist, "Consumers Gag on L.A.'s Toilet-to-Tap Program," *California Planning and Development Report*, June 1, 2000, www.cp-dr.com/articles/node-1278.

22. K. Chang, "The Water Cycle Gets Recycled: The East Valley Water Project Plans to Reuse Waste Water after Letting It Percolate through the Soil for Five Years," *Los Angeles Times*, December 5, 1995, www.latimes.com/archives/la-xpm-1995-12-05-me-10685-story.html.

23. Bill Van Wagoner, interview, October 8, 2019.

24. Sklar, "Toilet-to-Tap."

25. M. B. Haefele, "Daily Drips," *LA Weekly*, May 17, 2000.

26. Sklar, "Toilet-to-Tap."

27. M. B. Haefele and A. Sklar, "Revisiting 'Toilet to Tap,'" *Los Angeles Times*, August 26, 2007, www.latimes.com/opinion/la-op-haefele26aug26-story.html.

28. Ibid.

29. "Groundwater Replenishment," LADWP, accessed June 10, 2021, www.ladwp.com/ladwp/faces/ladwp/aboutus/a-water/a-w-recycledwater/a-w-rw-gwr?_adf.ctrl-state=19zylq0l73_4&_afrLoop=150434225488185.

30. "History," West Basin Municipal Water District, accessed June 10, 2021, www.westbasin.org/about-us/what-we-do/history.

31. "Recycled Water," West Basin Municipal Water District, accessed June 10, 2021, www.westbasin.org/water-supplies/recycled-water.

32. Sanitation Districts of Los Angeles County, *31st Annual Status Report on Recycled Water Use, FY 2019-20*, 1, 13, available at www.lacsd.org/water-reuse/annual_report.asp.

33. Ibid., 1.

34. "What is WIN?" Water Replenishment District of Southern California, accessed June 10, 2021, www.wrd.org/content/what-win.

35. Ted Johnson, interview, May 12, 2021.

36. J. Dill, "Demonstrating the Feasibility of Large-Scale Reuse in Southern California," *Municipal Water Leader* 6, no. 9 (2019): 12–15.

37. Ibid.

38. B. Apgar, "Nevada Could Get Some of California's Share of Lake Mead. Here's How," *Las Vegas Review-Journal*, February 6, 2021, www.reviewjournal.com/news/politics-and-government/nevada/nevada-could-get-some-of-californias-share-of-lake-mead-heres-how-2275031.

39. "Mayor Garcetti: Los Angeles Will Recycle 100% of City's Wastewater by 2035," *City of Los Angeles*, February 21, 2019, www.lamayor.org/mayor-garcetti-los-angeles-will-recycle-100-city's-wastewater-2035.

40. S. Catanzano, "State Must Analyze Practice of Dumping Billions of Gallons of Wastewater into Sea," *westsidetoday.com*, August 20, 2020, https://westsidetoday.com/2020/08/20/state-must-analyze-practice-of-dumping-billions-of-gallons-of-wastewater-into-sea.

41. California Constitution, Article X, Section 2.

42. Catanzano, "State Must Analyze Practice of Dumping."

43. Ibid.

44. E. Guerin, "LA Explained: The Los Angeles River," *LAist*, June 22, 2018, https://laist.com/2018/06/22/la_explained_the_la_river.php.

45. "About Us," Friends of the LA River, accessed June 10, 2021, https://folar.org/about-us.

46. G. Sencan and C. Chappelle, "The LA River and the Trade-Offs of Water Recycling," Public Policy Institute of California, June 24, 2019, www.ppic.org/blog/the-la-river-and-the-trade-offs-of-water-recycling.

47. L. Sahagun, "Lewis MacAdams, Famed Crusader for the Los Angeles River, Dies at 75," *Los Angeles Times*, April 21, 2020, www.latimes.com/story/2020-04-21/lewis-macadams-los-angeles-river-dies.

48. N. Wallet, "LA River Restoration Effort Lands $1.8 Million," *Los Angeles Daily News*, February 13, 2020, www.dailynews.com/2020/02/13/la-river-restoration-effort-lands-1-8-million.

49. "About Us," Friends of the LA River.

50. Earle Hartling, interview, October 8, 2019.

CHAPTER FOUR

1. I. Lobet, "Living on Earth: From Toilet to Tap," Public Radio International, January 18, 2008, www.loe.org/shows/segments.html?programID=08-P13-00003&segmentID=5.

2. K. J. Ormerod and L. Silvia, "Newspaper Coverage of Potable Water Recycling at Orange County Water District's Groundwater Replenishment System, 2000–2016," *Water* 9, no. 12 (2017): 984.

3. N. Masters, "When Orange County Was Rural (and Oranges Actually Grew There)," KCET, February 7, 2014, www.kcet.org/shows/lost-la/when-orange-county-was-rural-and-oranges-actually-grew-there.

4. Adam Hutchinson, written communication, January 28, 2020.

5. J. W. Moyer, "Jerry Brown Battles Calif. Water Crisis Created by His Father, Gov. Pat Brown," *Washington Post*, April 2, 2015, www.washington post.com/news/morning-mix/wp/2015/04/02/gov-jerry-brown-battling -california-water-crisis-created-by-his-father-gov-pat-brown.

6. "Water Factory 21," Orange County Water District, accessed June 10, 2021, www.ocwd.com/media/2451/water-factory-21-brochure.pdf.

7. Ibid.

8. Ibid.

9. The pilot project approaches included distillation (Texas), multi-stage flash distillation (San Diego), electrodialysis, vapor compression, and freezing. Reverse osmosis for seawater desalination was considered too expensive at the time.

10. *John F. Kennedy: Containing the Public Messages, Speeches, and Statements of the President, January 20 to December 31, 1961* (Washington, DC: U.S. Government Printing Office, 1962), 467.

11. T. Perry, "Carlsbad, Calif.'s $1 Billion Desalination Plant Touted as Largest in Western Hemisphere," Government Technology, June 5, 2015, www.govtech.com/fs/Carlsbad-Califs-1-Billion-Desalination-Plant-Touted-as -Largest-in-Western-Hemisphere.html.

12. In 1997, based on the results of an early survey, the project name was changed from Orange County Regional Water Reclamation Project to the Groundwater Replenishment System.

13. J. Schwartz, "Water Flowing from Toilet to Tap May Be Hard to Swallow," *New York Times*, May 8, 2015, www.nytimes.com/2015/05/12/science/recycled-drinking-water-getting-past-the-yuck-factor.html.

14. Orange County Water District and Orange County Sanitation District, GWRS: Groundwater Replenishment System, available at www.ocwd.com/media/8861/ocwd-technicalbrochure_web-2020.pdf.

15. S. Fakhreddine, J. Dittmar, D. Phipps, J. Dadakis, and S. Fendorf, "Geochemical Triggers of Arsenic Mobilization during Managed Aquifer Recharge," *Environmental Science & Technology* 49, no. 13 (2015): 7802–9.

16. "About GWRS," Orange County Water District, accessed June 10, 2021, www.ocwd.com/gwrs/about-gwrs.

17. Orange County Water District and Orange County Sanitation District, GWRS.

18. Ibid.

19. Ibid.

20. M. Lee, "S.D. Looks North for Help Marketing Recycled Water," *San Diego Union-Tribune*, September 12, 2005, www.sandiego.gov/sites/default/files/legacy/water/pdf/purewater/050912.pdf.

21. Lobet, "Living on Earth."

22. "Reclamation Project Makes Orange County 'Drought-Proof,'" *Los Angeles Daily News*, October 24, 2004, https://trib.com/news/state-and-regional/reclamation-project-makes-orange-county-drought-proof/article_79202204 -8d3f-5e00-af44-e75406a15a7c.html.

23. M. R. Markus and E. Torres, "How to Overcome Public Perception Issues on Potable Reuse Projects," accessed June 10, 2021, www.ocwd.com/media/7984/gwrs-outreach.pdf.

24. Ibid.

25. National Research Council, *Water Reuse: Potential for Expanding the Nation's Water Supply Through Reuse of Municipal Wastewater* (Washington, DC: National Academies Press, 2012), 66.

26. S. R. Harris-Lovett et al., "Beyond User Acceptance: A Legitimacy Framework for Potable Water Reuse in California," *Environmental Science & Technology* 49, no. 13 (2015): 7555.

27. NDMA precursors are compounds that form NDMA during disinfection by chlorine or chloramine.

28. G. Woodside and M. Westropp, "Orange County Water District Groundwater Management Plan 2015 Update," www.ocwd.com/media/2605/groundwatermanagementplan2015updatefinaldraft.pdf.

29. Ibid.

30. "Water Quality," Orange County Water District, accessed June 10, 2021, www.ocwd.com/what-we-do/water-quality.

31. Markus and Torres, "How to Overcome Public Perception Issues," 19.

CHAPTER FIVE

1. Cited in J. Rosenthal, "A Terrible Thing to Waste," *New York Times*, July 31, 2009, www.nytimes.com/2009/08/02/magazine/02FOB-onlanguage -t.html.

2. City of San Diego, Water Reuse Study (2006), accessed June 9, 2021, www.sandiego.gov/public-utilities/sustainability/pure-water-sd/reports/water -reuse-study.

3. M. Gonzalez, "Memo to San Diego Coastkeeper Board of Directors and Others, Re: Cooperative Agreement with City of San Diego for Non-Opposition to CWA 301(h) Waiver," January 29, 2009, available at https://groksurf.files .wordpress.com/2013/04/memo-to-activists-re-301h-with-exhibits.pdf.

4. Opened in December 2015, the Claude "Bud" Lewis Carlsbad Desalination Plant is the largest desalination plant in the Western Hemisphere, accounting for one-third of all water generated in San Diego County. Yet it meets less than 10 percent of the city's water demands.

5. California Department of Water Resources, Water Recycling 2030, Recommendations of California's Recycled Water Task Force (Sacramento: California Department of Water Resources, 2003), 21, https://cawaterlibrary .net/document/water-recycling-2030-recommendations-of-californias-recycled -water-task-force.

6. City of San Diego, Water Reuse Study, 1–3.

7. After allowances for treatment-process losses and other on-site uses, these plants have recycled-water production capacities of approximately 24 million gallons per day and 13.5 million gallons per day, respectively.

8. M. Lee, "Sanders Against Sending Treated Wastewater to Tap," *San Diego Union-Tribune*, July 20, 2006, www.sandiego.gov/sites/default/files/ legacy/water/pdf/purewater/060720.pdf.

9. The city council passed a resolution in 1989 mandating water recycling where feasible and otherwise beneficial, but this had little impact. See City of San Diego, Water Reuse Study, 4–12.

10. "Former Mayor Dick Murphy Wants Record Set Straight," *San Diego Union-Tribune*, October 8, 2011, www.sandiegouniontribune.com/sdut-former -mayor-dick-murphy-wants-record-set-straight-2011oct08-htmlstory.html.

11. Lee, "Sanders Against Sending."

12. Editorial Board, "Yuck! San Diego Should Flush 'Toilet to Tap' Plan," *San Diego Union-Tribune*, July 24, 2006, www.sandiego.gov/sites/default/ files/legacy/water/pdf/purewater/060724.pdf.

13. "Coalition Has Hand in Water Recycling Plan," *San Diego Union-Tribune*, January 27, 2010, www.sandiegouniontribune.com/sdut-coalition -has-hand-in-water-plan-ok-2010jan27-htmlstory.html; D. Lantry, "What's in the Water; What's in a Word: From Toilet-to-Tap to Pure Water," *Waterkeeper Magazine* 12, no. 1 (2016): 50–53.

14. Gonzalez, "Memo to San Diego Coastkeeper."

15. D. S. Brennan, "Time for Wastewater Turned Tap Water?" *San Diego Union-Tribune*, September 8, 2013, www.sandiegouniontribune.com/news/ environment/sdut-point-loma-wastewater-treatment-plant-waiver-2013sep08 -story.html.

16. Marsi Steirer, interview, November 18, 2019.

17. Gonzalez, "Memo to San Diego Coastkeeper."

18. Lantry, "What's in the Water."

19. F. Barringer, "As 'Yuck Factor' Subsides, Treated Wastewater Flows from Taps," *New York Times*, February 9, 2012, www.nytimes.com/2012/02/10/ science/earth/despite-yuck-factor-treated-wastewater-used-for-drinking.html.

20. L. Dillon, "City Water Usage Up," *Voice of San Diego*, February 9, 2010, www.voiceofsandiego.org/topics/government/city-water-usage-up.

21. R. Davis, "Will San Diego's Mayor Sip Purified Sewage?" *Voice of San Diego*, June 11, 2011, www.voiceofsandiego.org/topics/science-environment/will-san-diegos-mayor-sip-purified-sewage.

22. R. Davis, "Kittle's Departure Marks End of an Era," *Voice of San Diego*, August 12, 2009, www.voiceofsandiego.org/topics/economy/kittles-departure-marks-end-of-an-era.

23. Editorial Board, "The Yuck Factor: Get Over It," *San Diego Union-Tribune*, January 23, 2011, www.sandiego.gov/sites/default/files/legacy/water/purewater/pdf/2011/110123theyuckfactor.pdf.

24. The cost of imported water tripled in the past fifteen years and continues to rise. See "Pure Water San Diego, 2020 Year in Review Report," City of San Diego, accessed June 10, 2021, www.sandiego.gov/sites/default/files/pure_water_2020_year_in_review_pdf.pdf.

25. City of San Diego, Recycled Water Study (July 2012), ES-31, www.sandiego.gov/sites/default/files/legacy/water/pdf/purewater/2012/recycledfinaldraft120510.pdf.

26. Brennan, "Time for Wastewater."

27. S. Nunes, "Pure Water Initiative Passes City Council," KUSI News, November 19, 2014, www.kusi.com/pure-water-initiative-passes-city-council.

28. D. Garrick, "SD OKs Landmark Water Recycling," *San Diego Union-Tribune*, November 18, 2014, www.sandiegouniontribune.com/news/science/sdut-water-recycling-sewer-tap-council-approves-2014nov18-story.html.

29. Lantry, "What's in the Water."

30. "Cooperative Agreement in Support of Pure Water San Diego," October 2014, www.sandiego.gov/sites/default/files/cooperative_agreement_signed.pdf.

31. "Pure Water San Diego, 2020 Year in Review."

32. D. Garrick, "San Diego's Pure Water Sewage Recycling System Ready for Construction with All Hurdles Cleared," *San Diego Union-Tribune*, March 6, 2021, www.sandiegouniontribune.com/news/politics/story/2021-03-06/san-diegos-pure-water-sewage-recycling-system-ready-for-construction-after-litigation-delays.

33. "Pure Water Oceanside," City of Oceanside, accessed June 9, 2021, www.ci.oceanside.ca.us/gov/water/pure_water_oceanside.asp.

34. "East County Advanced Water Purification," accessed June 9, 2021, www.eastcountyawp.com.

35. The concentrate from reverse osmosis would still need to go to the Point Loma plant.

36. K. Meehan, K. J. Ormerod, and S. A. Moore, "Remaking Waste as Water: The Governance of Recycled Effluent for Potable Water Supply," *Water Alternatives* 6, no. 1 (2013): 67–85.

37. E. Yates, "East County Residents Speak Out against Proposed El Monte Valley Water Reclamation Project," *La Mesa Patch*, March 9, 2011, https://patch.com/california/lamesa/east-county-residents-speak-out-against -proposed-el-m010f9e016d.

38. "Helix Puts El Monte Water Project on Ice," *East County Magazine*, September 7, 2011, www.eastcountymagazine.org/helix-puts-el-monte-water -project-ice.

39. "East County Advanced Water Purification."

CHAPTER SIX

1. "Curse of the Boulder Valley," Wikipedia, accessed August 24, 2021, https://en.wikipedia.org/wiki/Curse_of_the_Boulder_Valley.

2. S. Hinchman, "Two Forks Proposal Has Roused Western Colorado," *High Country News* 20, no. 9 (1988): 11–12.

3. M. Tolchin, "E.P.A. to Veto Huge Colorado Dam," *New York Times*, November 24, 1990, 8; U.S. Environmental Protection Agency, *Recommended Determination to Prohibit Construction of Two Forks Dam and Reservoir Pursuant to Section 404(c) of the Clean Water Act* (Denver: Region VIII, March 1990).

4. S. Greene, "The Dance of the Dinosaurs," *Denver Post*, October 28, 2000, https://extras.denverpost.com/news/news1028d.htm.

5. S. Snyder, "Denver Water's Crucible: Two Forks," Denver Water, July 30, 2018, https://denverwatertap.org/2018/07/30/denver-waters-crucible-two -forks.

6. P. N. Limerick and J. L. Hanson, *A Ditch in Time* (Golden, CO: Fulcrum, 2012), 220.

7. D. A. Okun, "Water Reclamation and Unrestricted Nonpotable Reuse: A New Tool in Urban Water Management," *Annual Review of Public Health* 21 (2000): 226.

8. Denver Water, "Tracing Denver's Water History," July 11, 2018, https://denverwatertap.org/2018/07/11/tracing-denvers-water-history.

9. Limerick and Hanson, *A Ditch in Time*, 234.

10. S. Hall, "The Legacy of Colorado's Largest Wildfire," Denver Water, June 16, 2017, https://denverwatertap.org/2017/06/16/legacy-colorados -largest-wildfire. In 2020, three separate fires exceeded the acreage burned by the Hayman Fire.

11. Hall, "Colorado's Largest Wildfire."

12. Limerick and Hanson, *A Ditch in Time*, 65.

13. Hall, "Colorado's Largest Wildfire."

14. USDA Forest Service, *Water and the Forest Service* (Washington, DC: 2000).

15. *Colorado's Water Plan* (Denver: State of Colorado, 2015), xvii.

16. Ibid.

17. J. Crotty, "Building Lasting Relationships to Raise a Dam," *Municipal Water Leader* 4 (February 2018): 14–19.

18. Limerick and Hanson, *A Ditch in Time*, 168.

19. "Colorado River Cooperative Agreement (6-page summary)," Colorado River District, accessed May 30, 2021, www.coloradoriverdistrict.org/supply-planning/colorado-river-cooperative-agreement.

20. B. Udall, D. Kenney, and J. Fleck, "Western States Buy Time with a 7-Year Colorado River Drought Plan, But Face a Hotter, Drier Future," *The Conversation*, July 10, 2019, https://theconversation.com/western-states-buy-time-with-a-7-year-colorado-river-drought-plan-but-face-a-hotter-drier-future-119448.

21. *Colorado's Water Plan*, xxxi.

22. Ibid., xxx.

23. B. Finley, "Colorado Weighs Taking 'Waste' Out of Wastewater to Fix Shortfall," *Denver Post*, April 26, 2016, www.denverpost.com/2014/11/22/colorado-weighs-taking-waste-out-of-wastewater-to-fix-shortfall.

24. Limerick and Hanson, *A Ditch in Time*, 149.

25. W. Lorenz, "A History of Water Reuse in Colorado," Presentation at 28th Annual WateReuse Symposium, Denver, CO, September 15, 2013; Okun, "Water Reclamation and Unrestricted Nonpotable Reuse," 230.

26. A. Best, "Purified," *Headwaters* (Fall 2018): 15–24.

27. C. Carder, "Water, Water Not Everywhere," *Progressive Engineer*, accessed May 30, 2021, http://progressiveengineer.com/feature-water-water-not-everywhere.

28. W. C. Lauer et al., "Denver Potable Water Reuse Demonstration Project: Comprehensive Chronic Rat Study," *Food and Chemical Toxicology* 32, no. 11 (1994): 1021–30.

29. Finley, "Colorado Weighs."

30. Limerick and Hanson, *A Ditch in Time*, 25.

31. Ibid., 108.

32. "Residential: Xeriscape Plans," Denver Water, accessed June 17, 2021, www.denverwater.org/residential/rebates-and-conservation-tips/remodel-your-yard/xeriscape-plans.

33. A. Prendergast, "What's Killing the Trees in Denver Parks? It's the Water—and a Lot More," *Westword*, December 15, 2015, www.westword.com/news/whats-killing-the-trees-in-denver-parks-its-the-water-and-a-lot-more-7427509.

34. A. Parker, "Salinity Management for Landscapes," Presentation at 34th Annual WateReuse Symposium, San Diego, CA, September 10, 2019.

35. A. Prendergast, "Recycled Water Controversy: Denver Zoo Backs Off the Purple Pipe," *Westword*, December 23, 2015, www.westword.com/news/recycled-water-controversy-denver-zoo-backs-off-the-purple-pipe-7449703.

36. B. Finley, "Recycled Water Fight Stalls Plans for Bison, Birds, Fish and Museum," *Denver Post*, June 22, 2016, www.denverpost.com/2014/05/19/recycled-water-fight-stalls-plans-for-bison-birds-fish-and-museum.

37. Denver Water, "Leading the Way with the Denver Museum of Nature & Science," Mile High Water Talk (Blog), September 15, 2014, https://denverwater blog.org/2014/09/15/leading-the-way-with-the-denver-museum-of-nature -science/?iframe=true&preview=true/feed.

38. N. Gardner, "Aurora Is Growing Fast—and Isn't Slowing Down," *5280* (May 2019): 82–91.

39. Limerick and Hanson, *A Ditch in Time*, 135.

40. "Reclaimed Water," City of Aurora, accessed June 17, 2021, www.aurora gov.org/residents/water/water_system/water_treatment/reclaimed_water.

41. "Aurora Water History Timeline," Fact Sheet, February 22, 2012.

42. M. E. Sakas, "Aurora and Colorado Springs Want More Water. The Proposed Solution—A New Reservoir—Would Have Far-Reaching Impacts," *Colorado Public Radio*, March 18, 2021, www.cpr.org/2021/03/18/aurora -colorado-springs-water-new-reservoir-colorado-river.

43. J. Smith, "Aurora's Recycled Water Plant Running at Full-Tilt," *Water Education Colorado, Fresh Water News*, July 18, 2018, www.watereducation colorado.org/fresh-water-news/full-tilt-aurora-boosts-recycled-water-ops-90 -to-cope-with-this-summers-drought.

44. "Aurora Water Facts & Reports," accessed June 10, 2021, www.aurora gov.org/cms/One.aspx?portalId=16242704&pageId=16599815.

45. Ibid.

46. Greg Baker, interview, May 12, 2021.

47. Best, "Purified."

48. S. S. Paschke, ed., *Groundwater Availability of the Denver Basin Aquifer System, Colorado* (Reston, VA: U.S. Geological Survey, 2011).

49. Limerick and Hanson, *A Ditch in Time*, 230.

50. "History," South Metro Water Supply Authority, accessed June 17, 2021, https://southmetrowater.org/about/history.

51. L. Kilzer, J. Smith, and B. Hubbard, "Douglas Water Supply Sinking—Experts: Many Wells Could Be Useless in 10–20 Years," *Rocky Mountain News*, November 22, 2003, 1A.

52. B. Hubbard and J. Smith, "Warning Signs Ignored: 100-Year Water Rule Created by Panel Mistakenly Viewed as '100-year Supply,'" *Rocky Mountain News*, November 25, 2003, 5A.

53. In an unconfined aquifer, the principal source of water to a well is dewatering of the aquifer by drainage of the pores. In contrast, the water in confined aquifers is derived from aquifer compression and water expansion as the hydraulic pressure is reduced. If water levels in a confined aquifer are reduced to the point where it starts to become dewatered, the source of water becomes drainage of pores as in an unconfined aquifer.

54. Hubbard and Smith, "Warning Signs Ignored."

55. Ibid.

56. Ibid.

57. "Supply and Infrastructure," South Metro Water Supply Authority, accessed June 17, 2021, https://southmetrowater.org/our-work/supply -infrastructure.

58. "An Introduction to Rueter-Hess Reservoir," Parker Water and Sanitation District, accessed June 17, 2021, www.pwsd.org/2193/Rueter-Hess-Reservoir.

59. "Accomplishments," South Metro Water Supply Authority, accessed June 17, 2021, https://southmetrowater.org/our-plan/accomplishments.

60. Castle Rock Water, *Five-Year Strategic Plan, 2019–2023* (May 2019), 1.

61. "Imported Wise Water," Town of Castle Rock, accessed June 17, 2021, http://CRgov.com/WISEWater.

62. Mark Marlowe, interview, February 22, 2021.

63. Ibid.

64. "Reuse Water," Town of Castle Rock, accessed June 17, 2021, http:// crgov.com/3025/Reuse-Water.

CHAPTER SEVEN

1. B. Rankin, "Decades before Water Wars, Buford Dam Won City Support, Not Finances," *Atlanta Journal-Constitution*, August 10, 2012, www.ajc.com/news/ local/decades-before-water-wars-buford-dam-won-city-support-not-finances/ OVsYzGeeJOTLDSdjMkcGJL.

2. The far northeastern part of Georgia also includes Valley and Ridge and Appalachian Plateaus regions.

3. Georgia Environmental Protection Division, "Reuse Feasibility Analysis, EPD Guidance Document," August 2007, www1.gadnr.org/cws/Documents/ Reuse_Feasibility_Analysis.pdf.

4. Metropolitan North Georgia Water Planning District, "Water Resource Management Plan," June 2017, https://northgeorgiawater.org/wp-content/uploads/2015/05/Water-Resource-Management-Plan_Amended-20190227.pdf.

5. Rankin, "Decades before Water Wars."

6. Ibid.

7. W. M. Alley and R. Alley, *High and Dry: Meeting the Challenges of the World's Growing Dependence on Groundwater* (New Haven, CT: Yale University Press, 2017), 1.

8. "Tri-State Water Wars Overview," Atlanta Regional Commission, accessed June 10, 2021, https://atlantaregional.org/natural-resources/water/tri-state-water-wars-overview.

9. Rankin, "Decades before Water Wars."

10. E. M. Gilmer and J. Kay, "Supreme Court Hands Win to Georgia in State Water War (3)," *Bloomberg Law*, April 1, 2021, https://news.bloomberglaw.com/us-law-week/georgia-defeats-florida-in-supreme-court-water-war?mc_cid=a98e62a1ca&mc_eid=b76a339123.

11. "Water Plan Release Renews Regional Rivalries," *Gwinnett Daily Post*, June 23, 2007, www.gwinnettdailypost.com/archive/water-plan-release-renews-regional-rivalries/article_43c3ba67-0f82-517b-ab22-a771841ecf1e.html.

12. Wayne Hill, interview, April 17, 2020.

13. T. W. Hartley, "Public Perception and Participation in Water Reuse," *Desalination* 187 (2006): 122.

14. Ibid., 118.

15. "Critics Praise New Lake Lanier Water Discharge Plan," *Gwinnett Daily Post*, September 21, 2006, www.gwinnettdailypost.com/archive/critics-praise-new-lake-lanier-water-discharge-plan/article_ed6f4c3d-870d-55c5-88c5-8ac60b2515d1.html.

16. C. Young, "Wastewater Begins to Flow into Lanier," *Gwinnett Daily Post*, May 4, 2010, www.gwinnettdailypost.com/archive/wastewater-begins-to-flow-into-lanier/article_a25247a8-e7f4-5c4a-9bf7-9f7c57535a78.html.

17. At the front end, about one-third of the water goes through granular media filters and two-thirds ultrafiltration.

18. C. Yeomans, "Gwinnett Considered a Model for Treating, Reusing Water," *Gwinnett Daily Post*, May 8, 2015, www.gwinnettdailypost.com/archive/gwinnett-considered-a-model-for-treating-reusing-water-photos/article_f81b551f-522c-50cf-91bc-2525bab2ffa7.html.

19. "Critics Praise New Lake Lanier Water Discharge Plan."

20. U.S. Environmental Protection Agency, *Guidelines for Water Reuse* (Washington, DC: EPA/600/R-12/618, 2012), 3–30.

21. "F. Wayne Hill Water Resources Center," Gwinnett County, accessed June 10, 2021, www.gwinnettcounty.com/web/gwinnett/departments/water/whatwedo/wastewater/fwaynehillwaterresourcescenter.

22. Ibid.

23. "2018 Excellence in Environmental Engineering and Science Awards Competition Winner," American Academy of Environmental Engineers and Scientists, accessed June 10, 2021, www.aaees.org/e3scompetition/2018grandprize-research.php.

24. C. Yeomans, "Sculpture Honoring Former Gwinnett Chairman Wayne Hill Unveiled at EHC," *Gwinnett Daily Post*, December 18, 2018, www.gwinnettdailypost.com/local/sculpture-honoring-former-gwinnett-chairman-wayne-hill-unveiled-at-ehc/article_07b26483-4734-564a-b9a4-de6ad3c590fa.html.

25. "The Water Tower," accessed June 10, 2021, https://theh2otower.org.

26. National Research Council, *Water Reuse: Potential for Expanding the Nation's Water Supply Through Reuse of Municipal Wastewater* (Washington, DC: National Academies Press, 2012).

27. "Who We Are," Clayton County Water Authority, accessed June 10, 2021, www.ccwa.us/who-we-are.

28. D. Mayfield, "Could Your Sinks and Toilets Fight Sea-Level Rise in Hampton Roads?" *The Virginian-Pilot*, January 30, 2016, www.pilotonline.com/news/environment/article_df55a0e1-6992-54b0-b2c3-94a26402a89c.html.

29. T. Nading et al., "A 'SWIFT' Approach to Managed Aquifer Recharge," *Water Online*, January 24, 2018, www.wateronline.com/doc/a-swift-approach-to-managed-aquifer-recharge-0001.

30. J. Eggleston and J. Pope, *Land Subsidence and Relative Sea-Level Rise in the Southern Chesapeake Bay Region* (Reston, VA: U.S. Geological Survey, 2013).

31. R. Farley, "Climate Change Could Put the Navy's Norfolk Base Underwater," *Yahoo!news*, September 23, 2019, https://news.yahoo.com/climate-change-could-put-navys-035000628.html.

32. U.S. Environmental Protection Agency, "Consent Decree and Complaint: Hampton Roads," accessed June 10, 2021, www.epa.gov/enforcement/consent-decree-and-complaint-hampton-roads.

33. "Swift," Hampton Roads Sanitation District, accessed June 10, 2021, www.hrsd.com/swift.

34. J. Dill, "Hampton Roads Coastal Aquifer Recharge Program," *Municipal Water Leader* 6, no. 4 (2019): 10–13.

35. R. D. G. Pyne, *Aquifer Storage Recovery*, 2nd ed. (Gainesville, FL: ASR Press, 2005): 499–505.

36. U.S. Environmental Protection Agency, *Mainstreaming Potable Water Reuse in the United States: Strategies for Leveling the Playing Field* (ReNU-WIt and The Johnson Foundation at Wingspread, 2018), 12.

37. "US Water Alliance Announces US Water Prize 2018 Winners," accessed June 10, 2021, http://uswateralliance.org/news/us-water-alliance-announces-us-water-prize-2018-winners.

CHAPTER EIGHT

1. N. M. Blake, *Land into Water—Water into Land: A History of Water Management in Florida* (Tallahassee: University Press of Florida, 1980), 15.

2. "Honorary Members," WateReuse Association, accessed June 17, 2021, https://watereuse.org/about-watereuse/honorary-members; D. W. York, "Water Reuse: The Florida Model," accessed June 17, 2021, http://archives.water institute.ufl.edu/symposium/downloads/presentations/york.pdf.

3. Reuse Coordinating Committee and the Water Conservation Initiative Water Reuse Work Group, *Water Reuse for Florida: Strategies for Effective Use of Reclaimed Water* (Tallahassee: Florida Department of Environmental Protection, 2003), 33, https://floridadep.gov/sites/default/files/valued_resource_FinalReport_508C.pdf.

4. National Research Council, *Water Reuse: Potential for Expanding the Nation's Water Supply Through Reuse of Municipal Wastewater* (Washington, DC: National Academies Press, 2012), 34.

5. "Reclaimed Water," City of St. Petersburg, accessed June 17, 2021, www.stpete.org/water/water_services/reclaimed_water.php.

6. National Research Council, *Water Reuse*, 34.

7. "Reclaimed Water."

8. W. D. Johnson and J. R. Parnell, "Wastewater Reclamation and Reuse in the City of St. Petersburg, Florida," in *Wastewater Reclamation and Use*, ed. T. Asano (Lancaster, PA: Technomic, 1998), 1037–1104.

9. J. Crook, *Innovative Applications in Water Reuse: Ten Case Studies* (Alexandria, VA: WateReuse Association, 2004), 34–37.

10. Ibid.

11. Water Conserv II replaced Water Conserv I. The latter faced public opposition for its plan to inject high-quality reclaimed water into the upper layer of the Floridan aquifer. See "The Florida APRICOT Act of 1994," Florida Department of Environmental Protection, accessed June 17, 2021, https://floridadep.gov/water/domestic-wastewater/content/florida-apricot-act-1994.

12. "Welcome to Water CONSERV II," accessed June 17, 2021, www.waterconservii.com.

13. Reuse Coordinating Committee and the Water Conservation Initiative Water Reuse Work Group, *Water Reuse for Florida*, 2.

14. Florida Department of Environmental Protection, *2020 Reuse Inventory* (Tallahassee: Division of Water Resource Management, 2021), 2, https://floridadep.gov/water/domestic-wastewater/documents/2020-reuse -inventory-report.

15. C. Barnett, *Mirage: Florida and the Vanishing Water of the Eastern U.S.* (Ann Arbor: University of Michigan Press, 2007), 10.

16. J. D. Fretwell, J. S. Williams, and P. J. Redman, compilers, *National Water Summary on Wetland Resources* (Reston, VA: U.S. Geological Survey, 1996), 153.

17. Barnett, *Mirage*, 14–15.

18. R. L. Marella, *Water Withdrawals, Use, and Trends in Florida, 2015* (Reston, VA: U.S. Geological Survey, 2020).

19. F. H. Chapelle, *Wellsprings: A Natural History of Bottled Spring Waters* (New Brunswick, NJ: Rutgers University Press, 2005), 188.

20. T. M. Scott et al., *Springs of Florida* (Tallahassee: Florida Geological Survey, 2004).

21. "Glass Bottom Boats," Silver Springs State Park, accessed June 17, 2021, https://silversprings.com/glass-bottom-boats.

22. P. Burger and J. Kokjohn, "Wetlands to the Rescue—Recharging Our Water Supply," *Water Resources IMPACT* 22, no. 5 (2020): 34–36.

23. J. Staletovich, "Miami-Dade, Lagging Behind State, Seeks Cheap, Non-Icky Ways to Recycle Its Wastewater," *Miami Herald*, June 17, 2017, www .miamiherald.com/news/local/environment/article156659064.html.

24. Johnson and Parnell, "Wastewater Reclamation and Reuse in the City of St. Petersburg," 1040.

25. "Tampa Bay Water Wars," accessed June 17, 2021, http://hillsborough waterworks.com/water-wars.

26. Johnson and Parnell, "Wastewater Reclamation and Reuse," 1041.

27. Southwest Florida Water Management District, "50 Years," *WaterMatters Magazine* (October 2011), www.swfwmd.state.fl.us/sites/default/files/ medias/documents/50thAnniversary-WaterMatters.pdf.

28. R. Glennon, *Water Follies: Groundwater Pumping and the Fate of America's Fresh Waters* (Washington, DC: Island Press, 2002), 78.

29. "Tampa Bay Water Wars."

30. Glennon, *Water Follies*, 78.

31. Barnett, *Mirage*, 109.

32. H. Rand, *Water Wars: A Story of People, Politics and Power* (Philadelphia: Xlibris, 2003), 183.

33. "Groundwater," Tampa Bay Water, accessed June 17, 2021, www.tampa baywater.org/water-supply-source-groundwater.

34. "Reclaimed Water," Southwest Florida Water Management District, accessed June 17, 2021, www.swfwmd.state.fl.us/projects/reclaimed-water.

35. Reuse Coordinating Committee and the Water Conservation Initiative Water Reuse Work Group, *Water Reuse for Florida*, 102.

36. Johnson and Parnell, "Wastewater Reclamation and Reuse," 1045.

37. M. K. Hoppe, "Charting the Course for Tampa Bay's Recovery," *Bay Soundings*, December 3, 2019, http://baysoundings.com/charting-the-course-for-tampa-bays-recovery.

38. R. Danielson, "Tampa Looking at Long-Range Project to Turn Reclaimed Water into Drinking Water," *Tampa Bay Times*, April 19, 2015, www.tampabay.com/news/environment/water/tampa-looking-at-long-range-project-to-turn-reclaimed-water-into-drinking/2226174.

39. M. K. Hoppe, "Once Thought Impossible, Tampa Bay Leads Nation in Environmental Recovery," *Bay Soundings*, September 25, 2015, http://baysoundings.com/once-thought-impossible-tampa-bay-leads-nation-in-environmental-recovery.

40. J. Patterson, "Tampa Considers Plan to Add Waste Water to Drinking Water Supply," *WFLA*, July 26, 2019, www.wfla.com/news/local-news/tampa-considers-plan-to-add-waste-water-to-drinking-water-supply.

41. J. Zink, "Reclaimed Water Project Trickles to a Stop," *Tampa Bay Times*, August 28, 2005, www.tampabay.com/archive/2004/10/08/reclaimed-water-project-trickles-to-a-stop.

42. B. Baird et al., "Tampa Augmentation Project," *Florida Water Resources Journal* (April 2018): 12–18.

43. Danielson, "Tampa Looking at Long-Range Project."

44. R. Danielson, "Tampa Not Alone in Eyeing Highly Treated Waste Water as Possible Drinking Water Source," *Tampa Bay Times*, October 16, 2017, www.tampabay.com/news/environment/water/tampa-not-alone-in-eyeing-highly-treated-waste-water-as-possible-drinking/2340897.

45. "PURE: A Sustainable Water Alternative for Tampa," accessed June 17, 2021, www.tampa.gov/water/projects/pure.

46. "CS/SB 64—Reclaimed Water," the Florida Senate, accessed June 17, 2021, www.flsenate.gov/Committees/BillSummaries/2021/html/2320.

47. A. Bakst, "Tampa Is Thirsty for More Drinking Water," *WUSF*, December 3, 2019, https://wusfnews.wusf.usf.edu/post/tampa-thirsty-more-drinking-water.

48. Chuck Weber, interview, May 4, 2021.

49. Bart Weiss, interview, May 7, 2020.

50. V. Parsons, "Kracker Ave Restoration Thought to Be First of Its Kind, Former Fish Farm Property Offers Innovative Opportunities," *Bay Soundings*, February 6, 2019, http://baysoundings.com/kracker-restoration-thought-to-be-first-of-its-kind; Bart Weiss, interview, May 14, 2021.

51. S. London, "Florida County Aims for Full Usage of Reclaimed Water," *WaterWorld*, March 1, 2017, www.waterworld.com/wastewater/reuse-recycling/article/16191861/florida-county-aims-for-full-usage-of-reclaimed-water.

52. T. MacManus, "Once Leading the Way in Florida, Clearwater's Plan to Turn Wastewater to Drinking Water Is on Hold," *Tampa Bay Times*, January 18, 2019, www.tampabay.com/clearwater/once-leading-the-way-in-florida-clearwaters-plan-to-turn-wastewater-to-drinking-water-is-on-hold-20190122.

53. "Project Apricot," Altamonte Springs, accessed June 17, 2021, www.altamonte.org/953/Project-APRICOT.

54. "pureALTA," Altamonte Springs, accessed June 17, 2021, www.altamonte.org/754/pureALTA.

55. J. Kennedy, "'Gov. Poopy Water?' Groups Lobby Rick Scott to Veto Reused Water Bill," *Daytona Beach News-Journal*, March 28, 2018, www.news-journalonline.com/news/20180328/gov-poopy-water-groups-lobby-rick-scott-to-veto-reused-water-bill.

56. "CS/SB 64—Reclaimed Water."

57. Florida Potable Reuse Commission, *Framework for the Implementation of Potable Reuse in Florida* (Alexandria, VA: WateReuse Association, 2020).

CHAPTER NINE

1. N. Bowles, "Unfiltered Fervor: The Rush to Get Off the Water Grid," *New York Times*, December 29, 2017, www.nytimes.com/2017/12/29/dining/raw-water-unfiltered.html.

2. A. Hoyt, "'Raw Water' Is Having a Moment," How Stuff Works, February 1, 2018, https://health.howstuffworks.com/wellness/food-nutrition/facts/raw-water-is-having-moment.htm.

3. Ibid.

4. Bowles, "Unfiltered Fervor."

5. "Drinking Water," World Health Organization, accessed June 18, 2021, www.who.int/en/news-room/fact-sheets/detail/drinking-water.

6. There are cases where chemicals have caused short-term illness from drinking water. For example, more than three hundred cases were attributed to a large spill of a coal cleaning chemical into the Elk River in Charleston, West Virginia, in 2014. See W. M. Alley and R. Alley, *The War on the EPA: America's Endangered Environmental Protections* (Lanham, MD: Rowman & Littlefield, 2020), 175–77.

7. W. R. Mac Kenzie et al., "A Massive Outbreak in Milwaukee of Cryptosporidium Infection Transmitted through the Public Water Supply," *New England Journal of Medicine* 331, no. 3 (1994): 161–67; N. J. Hoxie et al.,

"Cryptosporidiosis-Associated Mortality Following a Massive Waterborne Outbreak in Milwaukee, Wisconsin," *American Journal of Public Health* 87, no. 12 (1997): 2032–35.

8. The EPA regulation is known as the Long-Term 2 Enhanced Surface Water Treatment Rule. Key provisions include source-water monitoring for Cryptosporidium, dual disinfectant inactivation for unfiltered systems, and potentially additional treatment for filtered systems based on Cryptosporidium concentrations in source water.

9. "Recycled Water Service," Eastern Municipal Water District, accessed June 18, 2021, www.emwd.org/recycled-water-service.

10. "EMWD Touts 'Groundwater Reliability Plus Initiative,'" *Patch*, February 21, 2019, https://patch.com/california/lakeelsinore-wildomar/emwd-touts-groundwater-reliability-plus-initiative.

11. California State Water Resources Control Board, A Proposed Framework for Regulating Direct Potable Reuse in California, 2nd ed. (Sacramento: California State Water Resources Control Board, 2019), 1, www.waterboards.ca.gov/drinking_water/certlic/drinkingwater/documents/direct_potable_reuse/dprframewkseced.pdf.

12. T. A. Johnson, "Groundwater Recharge Using Recycled Municipal Waste Water in Los Angeles County and the California Department of Public Health's Draft Regulations on Aquifer Retention Time," *Ground Water* 47, no. 4 (2009): 496–99.

13. A. W. Olivieri et al., Expert Panel Final Report: Evaluation of the Feasibility of Developing Uniform Water Recycling Criteria for Direct Potable Reuse (Sacramento: Prepared by the National Water Research Institute for the State Water Resources Control Board, 2016), 5.

14. Ibid.; B. Pecson et al., "Editorial Perspectives: Will SARS-CoV-2 Reset Public Health Requirements in the Water Industry? Integrating Lessons of the Past and Emerging Research," *Environmental Science: Water Research & Technology* 6, no. 7 (2020): 1761–64.

15. Pecson et al., "Will SARS-CoV-2 Reset Public Health Requirements."

16. M. Hellmér et al., "Detection of Pathogenic Viruses in Sewage: Virus and Norovirus Outbreaks Provided Early Warnings of Hepatitis," *Applied and Environmental Microbiology* 80, no. 21 (2014): 6771–81.

17. G. Medema et al., "Presence of SARS-Coronavirus-2 RNA in Sewage and Correlation with Reported COVID-19 Prevalence in the Early Stage of the Epidemic in the Netherlands, *Environmental Science & Technology Letters* 7, no. 7 (2020): 511–16.

18. D. V. Pulver, "As Some States Reopen, Studying Sewage Could Help Stop the Coronavirus Pandemic," *USA Today*, May 19, 2020, www.usatoday.com/story/news/investigations/2020/05/19/coronavirus-sewage-can-indicate-virus-spread-before-symptoms-appear/3107823001.

19. Bart Weiss, interview, May 7, 2020.

20. Safe yield as used in Arizona is a relatively outdated concept among groundwater scientists as it does not account for environmental effects. See W. M. Alley and S. A. Leake, "The Journey from Safe Yield to Sustainability," *Ground Water* 42, no. 1 (2004): 12–16.

21. W. M. Alley and R. Alley, *High and Dry: Meeting the Challenges of the World's Growing Dependence on Groundwater* (New Haven, CT: Yale University Press, 2017).

22. P. Dillon et al., "Sixty Years of Global Progress in Managed Aquifer Recharge," *Hydrogeology Journal* 27 (2019): 1–30.

23. C. Graf, "After 90 Years of Reusing Reclaimed Water in Arizona, What's in Store?" *Arizona Water Resource* 24, no. 4 (2016): 6.

24. S. B. Megdal, P. Dillon, and K. Seasholes, "Water Banks: Using Managed Aquifer Recharge to Meet Water Policy Objectives," *Water* 6 (2014): 1500–1514.

25. J. Robbins, "In Era of Drought, Phoenix Prepares for a Future Without Colorado River Water," *Yale Environment* 360, February 7, 2019, https://e360.yale.edu/features/how-phoenix-is-preparing-for-a-future-without-colorado-river-water.

26. "Recycled Water," City of Scottsdale, accessed June 18, 2021, www.scottsdaleaz.gov/water/recycled-water.

27. W. Tenney, "SROG—Wastewater Collaboration Through a Unique Partnership," AMWUA Blog, November 11, 2019, www.amwua.org/blog/srog--wastewater-collaboration-through-a-unique-partnership.

28. N. Sherbert, "Scottsdale Water Issued Arizona's First Permit for Direct Use of Recycled Water," *City of Scottsdale Water News*, September 13, 2019, www.scottsdaleaz.gov/water/news/scottsdale-water-issued-arizonas-first-permit-for-direct-use-of-recycled-water_s1_p28333.

CHAPTER TEN

1. Quote in J. Mervis, "Donald Kennedy," *Science* 370, no. 6523 (2020): 1401.

2. T. Heberer et al., "Detection of Drugs and Drug Metabolites in Ground Water Samples of a Drinking Water Treatment Plant," *Fresenius Environmental Bulletin* 6, no. 7–8 (1997): 438–43; T. A. Ternes, "Occurrence of Drugs in German Sewage Treatment Plants and Rivers," *Water Research* 32, no. 11 (1998): 3245–60.

3. D. W. Kolpin et al., "Pharmaceuticals, Hormones, and Other Wastewater Contaminants in U.S. Streams, 1999–2000: A National Reconnaissance," *Environmental Science & Technology* 36, no. 6 (2002): 1202–11.

4. S. Sauvé and M. Desrosiers, "A Review of What Is an Emerging Contaminant," *Chemistry Central Journal* 8 (2014): doi.org/10.1186/1752-153X-8-15.

5. J. Drewes et al., *Monitoring Strategies for Constituents of Emerging Concern (CECs) in Recycled Water* (Southern California Coastal Water Research Project, 2018), 52, https://ftp.sccwrp.org/pub/download/DOCUMENTS/TechnicalReports/1032_CECMonitoringInRecycledWater.pdf.

6. "Artificial Sweeteners and Cancer," National Cancer Institute, accessed June 18, 2021, www.cancer.gov/about-cancer/causes-prevention/risk/diet/artificial-sweeteners-fact-sheet.

7. Drewes et al., *Monitoring Strategies*, 33.

8. National Research Council, *Water Reuse: Potential for Expanding the Nation's Water Supply Through Reuse of Municipal Wastewater* (Washington, DC: National Academies Press, 2012), 66.

9. M. Brooks, "US Prescriptions Hit New High in 2018, but Opioid Scripts Dip," *Medscape*, May 10, 2019, www.medscape.com/viewarticle/912864.

10. "Contaminants of Emerging Concern," Water Quality Association, accessed June 18, 2021, www.wqa.org/whats-in-your-water/emerging-contaminants.

11. M. J. Benotti et al., "Pharmaceuticals and Endocrine Disrupting Compounds in U.S. Drinking Water," *Environmental Science & Technology* 43, no. 3 (2009): 597–603.

12. National Research Council, *Water Reuse*, 5.

13. R. Becker, "Premature or Precautionary? California Is First to Tackle Microplastics in Drinking Water," *CalMatters*, March 15, 2021, https://calmatters.org/environment/2021/03/california-microplastics-drinking-water.

14. A. Pruden, "Balancing Water Sustainability and Public Health Goals in the Face of Growing Concerns about Antibiotic Resistance," *Environmental Science & Technology* 48, no. 1 (2014): 5–14.

15. A. W. Olivieri et al., *Expert Panel Final Report: Evaluation of the Feasibility of Developing Uniform Water Recycling Criteria for Direct Potable Reuse* (Sacramento: Prepared by the National Water Research Institute for the State Water Resources Control Board, 2016), 185.

16. Drewes et al., *Monitoring Strategies*.

17. Ibid.

18. S. Harris-Lovett and D. Sedlak, "Protecting the Sewershed," *Science* 369, no. 6510 (2020): 1429–30.

19. M. Cone, "The Mystery of the Vanishing DDT in the Ocean Near Los Angeles," *Scientific American*, March 13, 2013, www.scientificamerican.com/article/the-mystery-of-the-vanishing-ddt-in-the-ocean-near-los-angeles.

20. M. McGough, "Man Pleads Guilty to Dumping Massive Amount of Polluted Water into Stockton Sewers," *Sacramento Bee*, February 11, 2020, www.sacbee.com/news/local/crime/article240185802.html.

21. National Water Research Institute, *Enhanced Source Control Recommendations for Direct Potable Reuse in California* (Fountain Valley: Report prepared for California State Water Resources Control Board, 2020), www.waterboards.ca.gov/drinking_water/certlic/drinkingwater/docs/dpr-esc-2020.pdf.

22. Orange County Sanitation District, *2019–2020 Pretreatment Program Annual Report*, available at www.ocsd.com/home/showpublisheddocument?id=30137.

23. M. Wisckol, "Orange County Water Districts Consider Massive Lawsuit Over PFAS Contamination," *Orange County Register*, May 22, 2020, www.ocregister.com/2020/05/22/orange-county-water-districts-consider-massive-lawsuit-over-pfas-contamination.

24. Orange County Water District, "PFAS in Orange County," Fact Sheet, November 2020, www.ocwd.com/media/9551/feb21_pfas-fact-sheet_final.pdf.

CHAPTER ELEVEN

1. M. Quartiano, *Toilet to Tap: San Diego's "Pipe Dream"* (self-published manuscript, 2006), 36, available at San Diego Public Library.

2. Ibid.

3. Ibid.

4. D. Sedlak, *Water 4.0: The Past, Present, and Future of the World's Most Vital Resource* (New Haven, CT: Yale University Press, 2014), 71.

5. Van Vuuren is quoted in P. L. Du Pisani, "Surviving in an Arid Land: Direct Reclamation of Potable Water at Windhoek's Goreangab Reclamation Plant," *Arid Lands Newsletter* 56 (November–December 2004).

6. R. E. Bichell, "As More Western Cities Turn to Recycled Water, They May Face a Curious Obstacle: The Ick Factor," *KUER 90.1*, June 29, 2018, www.kuer.org/post/more-western-cities-turn-recycled-water-they-may-face-curious-obstacle-ick-factor#stream/0.

7. P. Rozin et al., "Psychological Aspects of the Rejection of Recycled Water: Contamination, Purification and Disgust," *Judgment and Decision Making* 10, no. 1 (2015): 50–63.

8. Bichell, "As More Western Cities."

9. Calculation by geologist Robert Giegengack at the University of Pennsylvania, as reported by Rozin et al., "Psychological Aspects," 51.

10. P. Tennyson and M. Millan, "Toilet-to-Tap—Get Over It!" Presentation at 34th Annual WateReuse Symposium, San Diego, September 8–11, 2019.

11. D. Schultz, "EPA Water Reuse Plan Flush with Ideas Such as Toilet-to-Tap," *Bloomberg Environment*, September 10, 2019, https://news.bloomberg environment.com/environment-and-energy/water-reuse-plan-aims-to-boost -science-implementation-epa-says.

12. J. Molina, "Santa Barbara Renames Its El Estero Water Treatment Plant," *Noozhawk*, April 18, 2019, www.noozhawk.com/article/city _renames_el_estero_water_treatment_plant_santa_barbara.

13. "Get Over It? When It Comes to Recycled Water, Consumers Won't," *ScienceDaily*, November 18, 2019, www.sciencedaily.com/releases/ 2019/11/191118190851.htm.

14. S. R. Harris-Lovett et al., "Beyond User Acceptance: A Legitimacy Framework for Potable Water Reuse in California," *Environmental Science & Technology* 49, no. 13 (2015): 7552–61.

15. J. Allhands, "Yes, You Should Drink Beer Made from Recycled Water," *Arizona Republic*, September 11, 2017, www.azcentral.com/story/opinion/op-ed/ joannaallhands/2017/09/11/beer-made-recycled-water-taste-test/646426001.

16. J. Dill, "Turning Reuse Water into Beer: Pure Water Brew," *Municipal Water Leader* 6, no. 4 (2019): 22–25.

17. S. Dubois, "Brewers Tasked with Turning Sewage into Suds," *Associated Press*, April 28, 2015, www.sandiegouniontribune.com/sdut-brewers -tasked-with-turning-sewage-into-suds-2015apr28-story.html.

18. B. Nakamura, "Pure Water Brewing Alliance: The World's Most Sustainable Beer," accessed June 18, 2021, https://wef.org/globalassets/ assets-wef/3---resources/topics/o-z/water-reuse/reuse-beer/PWBA-StoneStory.

19. B. Chappell, "Beer Brewers Test a Taboo, Recycling Water after It Was Used in Homes," *NPR*, March 24, 2017, www.npr.org/sections/thetwo -way/2017/03/24/521388995/beer-brewers-test-a-taboo-recyling-water-after-it -was-used-in-homes.

20. Nakamura, "Pure Water Brewing Alliance."

21. C. Proctor, "Is Denver Water's 100th Anniversary Beer Any Good?" *Denver Water Tap*, July 20, 2018, https://denverwatertap.org/2018/07/20/ is-denver-waters-100th-anniversary-beer-any-good.

22. "Gov. John Hickenlooper Tells Senate Committee He Drank Fracking Fluid," *HuffPost*, February 14, 2013, www.huffpost.com/entry/ gov-john-hickenlooper-drank-fracking-fluid-hydraulic-fracturing_n_2674453.

23. "Pure Water Brewing Alliance: Water Should Be Judged by Its Quality, Not History," accessed June 18, 2021, www.wef.org/globalassets/assets-wef/3 ---resources/topics/o-z/water-reuse/reuse-beer/pwba-history; Dill, "Turning Reuse Water into Beer."

24. National Research Council, *Water Reuse: Potential for Expanding the Nation's Water Supply Through Reuse of Municipal Wastewater* (Washington, DC: National Academies Press, 2012), 102.

25. National Research Council, *Quality Criteria for Water Reuse* (Washington, DC: National Academy Press, 1982).

26. National Research Council, *Water Reuse*, 124.

27. B. M. Pecson et al., "Achieving Reliability in Potable Reuse: The Four Rs," *Journal AWWA* 107, no. 3 (2015): 48–58.

28. Ibid.

29. National Research Council, *Water Reuse*, 94–95.

30. Turbidity is a measure of the degree to which the water loses its transparency due to the presence of suspended particulates. The more total suspended solids in the water, the murkier it appears and the higher the turbidity.

31. E. Royte, "A Tall, Cool Drink of . . . Sewage?" *New York Times*, August 8, 2008, www.nytimes.com/2008/08/10/magazine/10wastewater-t.html.

32. The Expert Panel on Groundwater, *The Sustainable Management of Groundwater in Canada* (Ottawa: The Council of Canadian Academies, 2009); J. Brooke, "Few Left Untouched After Deadly *E. coli* Flows through an Ontario Town's Water," *New York Times*, July 10, 2000, A8.

33. Brooke, "Few Left Untouched."

34. F. Bajak, A. Suderman, and T. Lush, "Hack Exposes Vulnerability of Cash-Strapped US Water Plants," *AP News*, February 9, 2021, https://apnews .com/article/business-water-utilities-florida-coronavirus-pandemic-utilities -e783b0f1ca2af02f19f5a308d44e6abb.

35. "Recycled Water History," Otay Water District, accessed June 18, 2021, https://otaywater.gov/about-otay/water-information/reclamation-water/ recycled-water-history.

36. A. Krueger, "Merchants Told Water Is Tainted," *San Diego Union-Tribune*, August 22, 2007, A1, A12.

37. R. Davis, "A History of Death Threats, Scandal and Sewage-Tainted Water," *Voice of San Diego*, October 16, 2011, www.voiceofsandiego.org/ topics/science-environment/a-history-of-death-threats-scandal-and-sewage -tainted-water.

38. "Water Infrastructure," CSU Urban Water Center, accessed June 18, 2021, https://erams.com/urbanwatercenter/water-infrastructure.

39. P. C. Ingram et al., "From Controversy to Consensus: The Redwood City Recycled Water Experience," *Desalination* 187, no. 1–3 (2006): 179–90.

40. California Department of Water Resources, *Water Recycling 2030, Recommendations of California's Recycled Water Task Force* (Sacramento: California Department of Water Resources, 2003), G-79, https://cawate rlibrary.net/document/water-recycling-2030-recommendations-of-californias -recycled-water-task-force.

41. Ibid., G-80.

42. Mark Millan, personal communication, September 4, 2020.

43. "Recycled Water," City of Redwood City, accessed June 18, 2021, www .redwoodcity.org/departments/public-works/water/recycled-water.

44. Sedlak, *Water 4.0*, 198.

45. A. W. Olivieri et al., *Expert Panel Final Report: Evaluation of the Feasibility of Developing Uniform Water Recycling Criteria for Direct Potable Reuse* (Sacramento: Prepared by the National Water Research Institute for the State Water Resources Control Board, 2016), 15.

46. C. Fishman, *The Big Thirst: The Secret Life and Turbulent Future of Water* (New York: Free Press, 2011), 150.

47. Ibid., 154.

48. "Water Treatment," Seqwater, accessed June 18, 2021, www.seqwater .com.au/water-treatment; A. Hurlimann and S. Dolnicar, "When Public Opposition Defeats Alternative Water Projects—the Case of Toowoomba Australia," *Water Research* 44, no. 1 (2010): 287–97.

49. Hurlimann and Dolnicar, "When Public Opposition Defeats."

50. "Singapore Water Story," Public Utilities Board (PUB), accessed June 18, 2021, www.pub.gov.sg/watersupply/singaporewaterstory.

51. Sedlak, *Water 4.0*, 272.

CHAPTER TWELVE

1. WateReuse Colorado, "Advancing Direct Potable Reuse to Optimize Water Supplies and Meet Future Demands, Executive Summary," September 2018, https://watereuse.org/wp-content/uploads/2015/03/WRCO-DPR _Executive-Summary_Sep-2018.pdf.

2. M. E. Webber, *Thirst for Power* (New Haven, CT: Yale University Press, 2016), 106.

3. J. T. McCoy, "U.S. Water Recovery System on the International Space Station," in *2012 Guidelines for Water Reuse* (Washington, DC: U.S. Environmental Protection Agency, 2012), D152–53.

4. J. Anderson, "Life Support Upgrades Arrive at Station," March 9, 2020, www.nasa.gov/centers/marshall/news/releases/2020/life-support-upgrades -arrive-at-station-improve-reliability-for-moon-mars-missions.html.

5. H. L. Leverenz, G. Tchobanoglous, and T. Asano, "Direct Potable Reuse: A Future Imperative," *Journal of Water Reuse and Desalination* 1, no. 1 (2011): 2–10.

6. Ibid.

7. National Research Council, *Issues in Potable Reuse: The Viability of Augmenting Drinking Water Supplies with Reclaimed Water* (Washington, DC: National Academy Press, 1998), 2.

8. National Research Council, *Water Reuse: Potential for Expanding the Nation's Water Supply Through Reuse of Municipal Wastewater* (Washington, DC: National Academies Press, 2012), 54.

9. C. Binz, N. B. Razavian, and M. Kiparsky, "Of Dreamliners and Drinking Water: Developing Risk Regulation and a Safety Culture for Direct Potable Reuse," *Water Resources Management* 32 (2018): 511–25.

10. N. N. Taleb, *The Black Swan: The Impact of the Highly Improbable* (New York: Random House, 2007).

11. S. R. Harris-Lovett et al., "Beyond User Acceptance: A Legitimacy Framework for Potable Water Reuse in California," *Environmental Science & Technology* 49, no. 13 (2015): 7552–61.

12. Binz, Razavian, and Kiparsky, "Of Dreamliners and Drinking Water."

13. G. Tchobanoglous et al., *Framework for Direct Potable Reuse* (Alexandria, VA: WateReuse Research Foundation, 2015).

14. A. R. Parker and C. R. Lawrence, "Water Capture by a Desert Beetle," *Nature* 414, no. 6859 (2001): 33–34.

15. M. Qadir et al., "Research History and Functional Systems of Fog Water Harvesting," *Frontiers in Water* 3 (2021), https://doi.org/10.3389/frwa.2021.675269.

16. J. Crook, *Regulatory Aspects of Direct Potable Reuse in California* (Fountain Valley, CA: National Water Research Institute, 2010).

17. National Research Council, *Water Reuse*; I. B. Law, "Advanced Reuse—from Windhoek to Singapore and Beyond," *Water* 30, no. 5 (2003): 44–50.

18. Crook, *Regulatory Aspects of Direct Potable Reuse.*

19. D. F. Metzler et al., "Emergency Use of Reclaimed Water for Potable Supply at Chanute, Kan.," *Journal AWWA* 50, no. 8 (1958): 1021–60.

20. P. P. Livingston and R. R. Bennett, *Geology and Ground-Water Resources of the Big Spring Area, Texas* (Washington, DC: U.S. Geological Survey, 1944).

21. C. E. Scruggs, C. B. Pratesi, and J. R. Fleck, "Direct Potable Water Reuse in Five Arid Inland Communities: An Analysis of Factors Influencing Public Acceptance," *Journal of Environmental Planning and Management* 63, no. 8 (2019): 1470–1500.

22. Ibid.

23. D. Weissmann, "Texas Town Closes the Toilet-to-Tap Loop: Is This Our Future Water Supply?" *Marketplace*, January 6, 2014, www.marketplace.org/2014/01/06/texas-town-closes-toilet-tap-loop-our-future-water-supply.

24. "Reclaimed Water," Colorado River Municipal Water District, accessed June 19, 2021, www.crmwd.org/water-sources/reuse.

25. D. Weissmann, "Texas Town Closes."

26. E. Steinle-Darling, "Total Water Solutions—the Many Faces of DPR in Texas," *Journal AWWA* (March 2015): 16–20.

27. I. Dille, "A Place in the Sun," *Bicycling*, August 15, 2013, www.bicycling .com/rides/a20012577/hottern-hell-hundred; "Hotter'N Hell Hundred Bike Ride," accessed June 4, 2021, www.hh100.org.

28. D. Nix et al., "Tracking Wichita Falls' Path from DPR to IPR," *Opflow* 47, no. 2 (2021): 10–15.

29. Ibid.

30. Scruggs, Pratesi, and Fleck, "Reuse in Five Arid Inland Communities."

31. D. A. Muller and R. D. Price, *Ground-Water Availability in Texas— Estimates and Projections through 2030* (Austin: Texas Department of Water Resources, 1979).

32. N. Kounang, "El Paso to Drink Treated Sewage Water Due to Climate Change Drought," *CNN*, December 5, 2018, www.cnn.com/2018/11/30/health/ water-climate-change-el-paso/index.html.

33. A. Espinola, "How One Utility Won Public Support for Potable Reuse," American Water Works Association, January 21, 2016, www.awwa.org/ AWWA-Articles/how-one-utility-won-public-support-for-potable-reuse.

34. National Research Council, *Ground Water Recharge Using Waters of Impaired Quality* (Washington, DC: National Academy Press, 1994), 212.

35. National Research Council, *Water Reuse*, 46.

36. "Aquifer Recharge," El Paso Water, accessed June 19, 2021, www .epwater.org/our_water/water_resources/aquifer_recharge.

37. Kounang, "El Paso to Drink."

38. Espinola, "How One Utility Won."

39. "Rio Grande and Elephant Butte," El Paso Water, accessed June 19, 2021, www.epwater.org/our_water/water_resources/rio_grande_and_elephant_butte.

40. "Center for Inland Desalination Systems," accessed June 19, 2021, www .utep.edu/engineering/cids.

41. Espinola, "How One Utility Won."

42. "Who We Are," TecH2O Center, accessed June 19, 2021, www.tech2o .org/about_us/who_we_are.

43. "Our History of Conservation," El Paso Water, accessed June 19, 2021, www.epwater.org/conservation/Billions_of_Gallons_Saved.

44. "El Paso Water Leads in Water Reuse Technology," El Paso Water, April 12, 2019, www.epwater.org/about_us/newsroom/older_stories/ el_paso_water_leads_in_water_reuse_technology.

45. Gilbert Trejo and Christina Montoya, interview, February 3, 2021.

46. "El Paso Water Leads in Water Reuse Technology."

47. A. Espinola, "Potable Reuse Coming of Age," American Water Works Association, January 7, 2016, www.awwa.org/AWWA-Articles/potable-reuse-coming-of-age.

48. Kounang, "El Paso to Drink."

49. Wikipedia, "Cloudcroft, New Mexico."

50. "Pioneering Water Reuse in the Old West," *WaterWorld*, May 1, 2009, www.waterworld.com/international/desalination/article/16200553/pioneering-water-reuse-in-the-old-west.

51. Livingston Associates, *PURe Water Project Treatment Plant Completion—Engineering Design Summary*, July 13, 2020.

52. D. Venable, E. Livingston, and J. Vandegrift, "Village of Cloudcroft PURe Water Project—Direct Potable Reuse," in *2017 Potable Reuse Compendium* (Washington, DC: U.S. Environmental Protection Agency, 2017): A.4-1.

53. Ibid.

54. Scruggs, Pratesi, and Fleck, "Reuse in Five Arid Inland Communities."

55. National Water Research Institute, *Developing Proposed Direct Potable Reuse Operational Procedures and Guidelines for Cloudcroft, New Mexico* (Fountain Valley, CA: National Water Research Institute, 2015).

56. Scruggs, Pratesi, and Fleck, "Reuse in Five Arid Inland Communities."

57. Livingston Associates, *PURe Water Project*.

58. Scruggs, Pratesi, and Fleck, "Reuse in Five Arid Inland Communities."

59. C. E. Scruggs and B. M. Thomson, "Opportunities and Challenges for Direct Potable Water Reuse in Arid Inland Communities," *Journal of Water Resources Planning and Management* 143, no. 10 (2017): 04017064.

60. U.S. Department of the Interior, *Water 2025: Preventing Crises and Conflict in the West* (Washington, DC: U.S. Bureau of Reclamation, 2005).

61. C. E. Scruggs et al., "Potable Water Reuse in Small Inland Communities: Oasis or Mirage?" *Journal AWWA* 112, no. 4 (2020): 10–17.

62. Tchobanoglous et al., *Framework for Direct Potable Reuse*.

63. A. W. Olivieri et al., *Expert Panel Final Report: Evaluation of the Feasibility of Developing Uniform Water Recycling Criteria for Direct Potable Reuse* (Sacramento, CA: Prepared by the National Water Research Institute for the State Water Resources Control Board, 2016).

64. "A Proposed Framework of Regulating Direct Potable Reuse in California Addendum Version 3-22-2021," www.waterboards.ca.gov/drinking_water/certlic/drinkingwater/documents/direct_potable_reuse/dprframewkaddendum.pdf.

CHAPTER THIRTEEN

1. Y. Berra, *The Yogi Book: "I Really Didn't Say Everything I Said"* (New York: Workman Publishing, 1998), 118–19.

2. "Gillette Stadium and Patriot Place," Natural Systems Utilities, accessed June 20, 2021, www.nsuwater.com/case-studies/gillette-stadium.

3. J. Dill, "Pentair: Water Reuse from Stadiums to Factories to Private Homes," *Municipal Water Leader* 6, no. 4 (2019): 16–19.

4. T. Hartman, "Big Water Savings Come Home in Groundbreaking Pilot Project," Denver Water Tap, May 14, 2020, https://denverwatertap .org/2020/05/14/big-water-savings-come-home-in-groundbreaking-pilot -project.

5. "Salesforce Announces Innovative Water Recycling System for Sales-force Tower," *Salesforce News and Insights*, January 11, 2018, www.salesforce .com/news/press-releases/2018/01/11/salesforce-announces-innovative-water -recycling-system-for-salesforce-tower.

6. N. Dornak, "One Water, One Amazing School: Texas Elementary School Embodies Water Conservation Inside and Out," Green Schools National Network, December 22, 2020, https://greenschoolsnationalnetwork .org/one-water-one-amazing-school-texas-elementary-school-embodies-water -conservation-inside-and-out.

7. San Francisco Public Utilities Commission, "Blueprint for Onsite Water Systems," September 2014, www.sfwater.org/modules/showdocument .aspx?documentid=6057.

8. National Academies of Sciences, Engineering, and Medicine, *Using Graywater and Stormwater to Enhance Local Water Supplies: An Assessment of Risks, Costs, and Benefits* (Washington, DC: National Academies Press, 2016).

9. S. Sharvelle et al., *Long-Term Study on Landscape Irrigation Using Household Graywater—Experimental Study* (Fort Collins, CO: Report to Water Environment & Reuse Foundation, 2012).

10. "Greywater Action," accessed June 20, 2021, https://greywateraction.org.

11. National Academies of Sciences, Engineering, and Medicine, *Using Graywater and Stormwater*.

12. K. Brulliard and W. Wan, "Put a Lid on It, Folks: Flushing May Release Coronavirus-Containing 'Toilet Plumes,'" *Washington Post*, June 16, 2020, www.washingtonpost.com/health/2020/06/16/coronavirus-toilet-flushing.

13. Z. L. T. Yu et al., "Critical Review: Regulatory Incentives and Impedi-ments for Onsite Graywater Reuse in the United States," *Water Environment Research* 85, no. 7 (2013): 650–62.

14. William J. Worthen Foundation, *A Design Professional's Guide to Onsite Water Use and Reuse* (San Francisco: William J. Worthen Foundation, 2018).

15. S. Sharvelle et al., *Risk-Based Framework for the Development of Public Health Guidance for Decentralized Non-Potable Water Systems* (Alexandria, VA: Prepared by the National Water Research Institute for the Water Environment & Reuse Foundation, 2017), 6.

16. H. Cooley et al., *The Role of Onsite Water Systems in Advancing Water Resilience in Silicon Valley* (Oakland, CA: Pacific Institute, 2021).

17. N. Hundley, *The Great Thirst: Californians and Water: A History* (Berkeley: University of California Press, 2001), 171.

18. Ibid., 186.

19. S. Craig, "Hetch Hetchy Water's Epic Journey, from Mountains to Tap," *KQED*, July 12, 2018, www.kqed.org/news/11674188/hetch-hetchy-waters-epic-journey-from-mountains-to-tap.

20. "Groundwater," San Francisco Public Utilities Commission, accessed June 20, 2021, https://sfpuc.org/programs/water-supply-planning/groundwater.

21. C. G. Hyde, "The Beautification and Irrigation of Golden Gate Park with Activated Sludge Effluent," *Sewage Works Journal* 9, no. 6 (1937): 929–41.

22. N. Pinhey, "Forgotten Facilities: Golden Gate Park's 1932 Recycled Water Plant," *CWEA*, November 23, 2019, http://cweawaternews.org/forgotten-facilities-golden-gate-parks-1932-recycled-water-plant.

23. "Recycled Water," San Francisco Public Utilities Commission, accessed June 20, 2021, https://sfpuc.org/programs/water-supply-planning/recycled-water.

24. "Living Machine," San Francisco Public Utilities Commission, accessed June 20, 2021, https://sfwater.org/index.aspx?page=1156.

25. San Francisco Public Utilities Commission, *Onsite Water Reuse Program Guidebook* (July 2020), https://sfpuc.org/sites/default/files/documents/OnsiteWaterReuseGuidebook2020.pdf.

26. Bay City News, "City Ordinance Seeks to Conserve Water Supply, Expand Recycled Water Amid Statewide Drought Conditions," *San Francisco Examiner*, May 12, 2021, www.sfexaminer.com/news/city-ordinance-seeks-to-conserve-water-supply-expand-recycled-water-amid-statewide-drought-conditions.

27. "Graywater/Laundry to Landscape," San Francisco Public Utilities Commission, accessed June 20, 2021, https://sfpuc.org/learning/conserve-water/graywater-laundry-landscape.

28. W. B. DeOreo et al., *Residential End Uses of Water, Version 2* (Denver: Water Research Foundation, 2016).

29. "PureWaterSF," San Francisco Public Utilities Commission, accessed June 20, 2021, www.sfpuc.org/programs/future-water-supply-planning/innovations/purewatersf.

30. San Francisco Public Utilities Commission, "Blueprint for Onsite Water Systems."

31. Sharvelle et al., *Risk-Based Framework*.

32. U.S. Water Alliance and Water Environment & Reuse Foundation, "National Blue Ribbon Commission for Onsite Nonpotable Water Systems," *Press Release*, December 14, 2016, https://sfwater.org/Modules/ShowDocumen.aspx?documentID=10195.

33. San Francisco Public Utilities Commission, "Blueprint for Onsite Water Systems."

34. C. Koehler and C. Koch, *Innovation in Action: 21st Century Water Infrastructure Solutions* (San Francisco: WaterNow Alliance, 2019), 64–68.

35. D. Sedlak, *Water 4.0: The Past, Present, and Future of the World's Most Vital Resource* (New Haven, CT: Yale University Press, 2014).

36. U.S. Environmental Protection Agency, *Clean Watersheds Needs Survey 2012 Report to Congress* (Washington, DC: U.S. Environmental Protection Agency, 2016).

37. J. G. Hering et al., "A Changing Framework for Urban Water Systems," *Environmental Science & Technology* 47 (2013): 10721–725.

38. William J. Worthen Foundation, *A Design Professional's Guide*.

39. C. Shin et al., "Pilot-Scale Temperate-Climate Treatment of Domestic Wastewater with a Staged Anaerobic Fluidized Membrane Bioreactor (SAF-MBR)," *Bioresource Technology* 159 (2014): 95–103.

40. "Codiga Resource Recovery Center Process Overview," accessed June 20, 2021, https://cr2c.stanford.edu/about-0.

CHAPTER 14

1. D. Sedlak, *Water 4.0: The Past, Present, and Future of the World's Most Vital Resource* (New Haven, CT: Yale University Press, 2014), 216.

2. J. Fulcher, "Changing the Terms," *WEF Highlights*, May 22, 2014, https://news.wef.org/changing-the-terms.

3. US Water Alliance, *One Water for America Policy Framework* (Washington, DC: US Water Alliance, 2017).

4. B. C. Gile et al., "Integrated Water Management at the Peri-Urban Interface: A Case Study of Monterey, California," *Water* 12, no. 12 (2020), https://doi.org/10.3390/w12123585.

5. "Pure Water Monterey," accessed June 20, 2021, www.montereyone water.org/261/Pure-Water-Monterey-Overview.

6. U.S. Environmental Protection Agency, *National Water Reuse Action Plan Draft* (Washington, DC: U.S. Environmental Protection Agency, 2019), 6, available at www.epa.gov/sites/production/files/2019-09/documents/water -reuse-action-plan-draft-2019.pdf.

7. Withdrawals for public supply are estimated to average about 39 billion gallons per day. See C. A. Dieter et al., *Estimated Use of Water in the United States in 2015* (Reston, VA: U.S. Geological Survey, 2018).

8. "Geysers Recharge," City of Santa Rosa, accessed June 20, 2021, https://srcity.org/3544/Geysers-Recharge.

9. D. Lach, H. Ingram, and S. Rayner, "Maintaining the Status Quo: How Institutional Norms and Practices Create Conservative Water Organizations," *Texas Law Review* 83, no. 7 (2005): 2027–53.

10. J. G. Hering et al., "A Changing Framework for Urban Water Systems," *Environmental Science & Technology* 47, no. 19 (2013): 10721–26.

11. G. T. Mehan III, "Found Water: Reuse and the Deconstruction of "Wastewater," *Water Finance & Management*, December 10, 2019.

12. U.S. Environmental Protection Agency, *National Water Reuse Action Plan: Collaborative Implementation (Version 1)* (Washington, DC: U.S. Environmental Protection Agency, 2020), available at www.epa.gov/waterreuse/ national-water-reuse-action-plan-collaborative-implementation-version-1.

13. For a discussion of the history and effectiveness of the Toxic Substances and Control Act and PFAS, see W. M. Alley and R. Alley, *The War on the EPA: America's Endangered Environmental Protections* (Lanham, MD: Rowman & Littlefield, 2020): 157–95.

14. L. A. Patterson and M. W. Doyle, *2020 Aspen-Nicholas Water Forum Water Affordability and Equity Briefing Document* (Durham, NC: Nicholas Institute, 2020).

15. M. P. Teodoro and R. R. Saywitz, "Water and Sewer Affordability in the United States: A 2019 Update," *AWWA Water Science* 2, no. 2 (2020): e1176.

16. B. DeOreo et al., *Residential End Uses of Water, Version 2: Executive Report* (Denver: Water Research Foundation, 2016).

17. B. D. Richter et al., "Decoupling Urban Water Use and Growth in Response to Water Scarcity," *Water* 12, no. 10 (2020), www.mdpi .com/2073-4441/12/10/2868.

18. K. Schwabe et al., "Unintended Consequences of Water Conservation on the Use of Treated Municipal Wastewater," *Nature Sustainability* 3 (2020): 628–35.

19. Gilbert Trejo, interview, February 3, 2021.

Selected Bibliography

Alley, W. M., and R. Alley. *High and Dry: Meeting the Challenges of the World's Growing Dependence on Groundwater*. New Haven, CT: Yale University Press, 2017.

———. *The War on the EPA: America's Endangered Environmental Protections*. Lanham, MD: Rowman & Littlefield, 2020.

Alley, W. M., and S. A. Leake. "The Journey from Safe Yield to Sustainability." *Ground Water* 42, no. 1 (2004): 12–16.

Asano, T., ed. *Wastewater Reclamation and Use*. Lancaster, PA: Technomic, 1998.

Baird, B., C. Weber, S. Park, D. Ammerman, and S. Burns. "Tampa Augmentation Project." *Florida Water Resources Journal* (April 2018): 12–18.

Barnett, C. *Mirage: Florida and the Vanishing Water of the Eastern U.S.* Ann Arbor: University of Michigan Press, 2007.

Benotti, M. J., R. A. Trenholm, B. J. Vanderford, J. C. Holady, B. D. Stanford, and S. A. Snyder. "Pharmaceuticals and Endocrine Disrupting Compounds in U.S. Drinking Water." *Environmental Science & Technology* 43, no. 3 (2009): 597–603.

Best, A. "Purified." *Headwaters* (Fall 2018): 15–24.

Binz, C., N. B. Razavian, and M. Kiparsky. "Of Dreamliners and Drinking Water: Developing Risk Regulation and a Safety Culture for Direct Potable Reuse." *Water Resources Management* 32 (2018): 511–25.

Blake, N. M. *Land into Water—Water into Land: A History of Water Management in Florida*. Tallahassee: University Press of Florida, 1980.

Burger, P., and J. Kokjohn. "Wetlands to the Rescue—Recharging Our Water Supply." *Water Resources IMPACT* 22, no. 5 (2020): 34–36.

California Department of Water Resources. *Water Recycling 2030, Recommendations of California's Recycled Water Task Force*. Sacramento: California Department of Water Resources, 2003. Available at https://cawaterlibrary.net/document/water-recycling-2030-recommendations-of-californias-recycled-water-task-force.

California State Water Resources Control Board. *A Proposed Framework for Regulating Direct Potable Reuse in California*, 2nd ed. Sacramento: California State Water Resources Control Board, 2019. Available at www.waterboards.ca.gov/drinking_water/certlic/drinkingwater/documents/direct_potable_reuse/dprframewkseced.pdf.

Chapelle, F. H. *Wellsprings: A Natural History of Bottled Spring Waters*. New Brunswick, NJ: Rutgers University Press, 2005.

City of San Diego. *Recycled Water Study*. San Diego: City of San Diego, 2012. Available at www.sandiego.gov/sites/default/files/legacy/water/pdf/purewater/2012/recycledfinaldraft120510.pdf.

———. *Water Reuse Study*. San Diego: City of San Diego, 2006. Available at www.sandiego.gov/public-utilities/sustainability/pure-water-sd/reports/water-reuse-study.

———. *Water Reuse Study: Water Reuse Goals, Opportunities & Values*. San Diego: City of San Diego, 2004. Available at www.sandiego.gov/sites/default/files/legacy/water/pdf/purewater/aa1wp.pdf.

Colorado's Water Plan. Denver: State of Colorado, 2015.

Cooley, H., A. Thebo, C. Kammeyer, and D. Bostic. *The Role of Onsite Water Systems in Advancing Water Resilience in Silicon Valley*. Oakland, CA: Pacific Institute, 2021.

Crook, J. *Innovative Applications in Water Reuse: Ten Case Studies*. Alexandria, VA: WateReuse Association, 2004.

———. *Regulatory Aspects of Direct Potable Reuse in California*. Fountain Valley: National Water Research Institute, 2010.

Crotty, J. "Building Lasting Relationships to Raise a Dam." *Municipal Water Leader* 4 (February 2018): 14–19.

DeOreo, W. B., P. Mayer, B. Dziegielewski, and J. Kiefer. *Residential End Uses of Water, Version 2*. Denver: Water Research Foundation, 2016.

Dieter, C. A., M. A. Maupin, R. R. Caldwell, M. A. Harris, T. I. Ivahnenko, J. K. Lovelace, N. L. Barber, and K. S. Linsey. *Estimated Use of Water in the United States in 2015*. Reston, VA: U.S. Geological Survey, 2018.

Dill, J. "Demonstrating the Feasibility of Large-Scale Reuse in Southern California." *Municipal Water Leader* 6, no. 9 (2019): 12–15.

———. "Hampton Roads Coastal Aquifer Recharge Program." *Municipal Water Leader* 6, no. 4 (2019): 10–13.

———. "Pentair: Water Reuse from Stadiums to Factories to Private Homes." *Municipal Water Leader* 6, no. 10 (2019): 16–19.

————. "Turning Reuse Water into Beer: Pure Water Brew." *Municipal Water Leader* 6, no. 4 (2019): 22–25.

Dillon, P., P. Stuyfzand, T. Grischek, M. Lluria, R. D. G. Pyne, R. C. Jain, J. Bear, et al. "Sixty Years of Global Progress in Managed Aquifer Recharge." *Hydrogeology Journal* 27 (2019): 1–30.

Drewes, J., P. Anderson, N. Denslow, W. Jakubowski, A. Olivieri, D. Schlenk, and S. Snyder. *Monitoring Strategies for Constituents of Emerging Concern (CECs) in Recycled Water*. Southern California Coastal Water Research Project, 2018. Available at https://ftp.sccwrp.org/pub/download/DOCUMENTS/TechnicalReports/1032_CECMonitoringInRecycledWater.pdf.

Du Pisani, P. L. "Surviving in an Arid Land: Direct Reclamation of Potable Water at Windhoek's Goreangab Reclamation Plant." *Arid Lands Newsletter* 56 (November–December 2004).

Eggleston, J., and J. Pope. *Land Subsidence and Relative Sea-Level Rise in the Southern Chesapeake Bay Region*. Reston, VA: U.S. Geological Survey, 2013.

Engineering-Science. *Monterey Wastewater Reclamation Study for Agriculture—Final Report*. Monterey, CA: Monterey Regional Water Pollution Control Agency, 1987.

The Expert Panel on Groundwater. *The Sustainable Management of Groundwater in Canada*. Ottawa: The Council of Canadian Academies, 2009.

Fakhreddine, S., J. Dittmar, D. Phipps, J. Dadakis, and S. Fendorf. "Geochemical Triggers of Arsenic Mobilization during Managed Aquifer Recharge." *Environmental Science & Technology* 49, no. 13 (2015): 7802–9.

Fishman, C. *The Big Thirst: The Secret Life and Turbulent Future of Water*. New York: Free Press, 2011.

Florida Department of Environmental Protection. *2020 Reuse Inventory*. Tallahassee: Division of Water Resource Management, 2021. Available at https://floridadep.gov/water/domestic-wastewater/documents/2020-reuse-inventory-report.

Florida Potable Reuse Commission. *Framework for the Implementation of Potable Reuse in Florida*. Alexandria, VA: WateReuse Association, 2020.

Fretwell, J. D., J. S. Williams, and P. J. Redman, compilers. *National Water Summary on Wetland Resources*. Reston, VA: U.S. Geological Survey, 1996.

Georgia Environmental Protection Division. *Reuse Feasibility Analysis, EPD Guidance Document*. Atlanta: Georgia Environmental Protection Division, August 2007. Available at www1.gadnr.org/cws/Documents/Reuse_Feasibility_Analysis.pdf.

Gile, B. C., P. A. Sciuto, N. Ashoori, and R. G. Luthy. "Integrated Water Management at the Peri-Urban Interface: A Case Study of Monterey, California." *Water* 12, no. 12 (2020): 3585. https://doi.org/10.3390/w12123585.

Glennon, R. *Unquenchable: America's Water Crisis and What to Do about It.* Washington, DC: Island Press, 2009.

Graf, C. "After 90 Years of Reusing Reclaimed Water in Arizona, What's in Store?" *Arizona Water Resource* 24, no. 4 (2016): 6.

Green, D. *Managing Water: Avoiding Crisis in California.* Berkeley: University of California Press, 2007.

Harris-Lovett, S., C. Binz, D. L. Sedlak, M. Kiparsky, and B. Truffer. "Beyond User Acceptance: A Legitimacy Framework for Potable Water Reuse in California." *Environmental Science & Technology* 49, no. 13 (2015): 7552–61.

Harris-Lovett, S., and D. Sedlak. "The History of Water Reuse in California." In *Sustainable Water*, edited by A. Lassiter, 220–43. Berkeley: University of California Press, 2015.

———. "Protecting the Sewershed." *Science* 369, no. 6510 (2020): 1429–30.

Hartley, T. W. "Public Perception and Participation in Water Reuse." *Desalination* 187, no. 1–3 (2006): 115–26.

Heberer, T., U. Dunnbier, C. Reilich, and H. Stan. "Detection of Drugs and Drug Metabolites in Ground Water Samples of a Drinking Water Treatment Plant." *Fresenius Environmental Bulletin* 6, no. 7–8 (1997): 438–43.

Hellmér, M., N. Paxéus, L. Magnius, L. Enache, B. Arnholm, A. Johansson, L. Bergström, and H. Norder. "Detection of Pathogenic Viruses in Sewage: Virus and Norovirus Outbreaks Provided Early Warnings of Hepatitis." *Applied and Environmental Microbiology* 80, no. 21 (2014): 6771–81.

Hering, J. G., T. D. Waite, R. G. Luthy, J. E. Drewes, and D. L. Sedlak. "A Changing Framework for Urban Water Systems." *Environmental Science & Technology* 47 (2013): 10721–725.

Hoxie, N. J., J. P. Davis, J. M. Vergeront, R. D. Nashold, and K. A. Blair. "Cryptosporidiosis-Associated Mortality Following a Massive Waterborne Outbreak in Milwaukee, Wisconsin." *American Journal of Public Health* 87, no. 12 (1997): 2032–35.

Hundley, N. *The Great Thirst: Californians and Water: A History.* Berkeley: University of California Press, 2001.

Hurlimann, A., and S. Dolnicar. "When Public Opposition Defeats Alternative Water Projects—the Case of Toowoomba Australia." *Water Research* 44, no. 1 (2010): 287–97.

Hyde, C. G. "The Beautification and Irrigation of Golden Gate Park with Activated Sludge Effluent." *Sewage Works Journal* 9, no. 6 (1937): 929–41.

Ingram, P. C., V. J. Young, M. Millan, C. Chang, and T. Tabucchi. "From Controversy to Consensus: The Redwood City Recycled Water Experience." *Desalination* 187, no. 1–3 (2006): 179–90.

Jaramillo, M. F., and I. Restrepo. "Wastewater Reuse in Agriculture: A Review about Its Limitations and Benefits." *Sustainability* 9, no. 10 (2017): 1734. doi.org/10.3390/su9101734.

Johnson, T. A. "Groundwater Recharge Using Recycled Municipal Waste Water in Los Angeles County and the California Department of Public Health's Draft Regulations on Aquifer Retention Time." *Ground Water* 47, no. 4 (2009): 496–99.

Katz, S. M., and P. Tennyson. "Coming Full Circle: Craft Brewers Demonstrate Potable Reuse Acceptance." *Journal AWWA* 110, no. 1 (2018): 62–67.

Koehler, C., and C. Koch. *Innovation in Action: 21st-Century Water Infrastructure Solutions*. San Francisco: WaterNow Alliance, 2019.

Kolpin, D. W., E. T. Furlong, M. T. Meyer, E. M. Thurman, S. D. Zaugg, L. B. Barber, and H. T. Buxton. "Pharmaceuticals, Hormones, and Other Wastewater Contaminants in U.S. Streams, 1999–2000: A National Reconnaissance." *Environmental Science & Technology* 36, no. 6 (2002): 1202–11.

Lach, D., H. Ingram, and S. Rayner. "Maintaining the Status Quo: How Institutional Norms and Practices Create Conservative Water Organizations." *Texas Law Review* 83, no. 7 (2005): 2027–53.

Lantry, D. "What's in the Water; What's in a Word: From Toilet-to-Tap to Pure Water." *Waterkeeper Magazine* 12, no. 1 (2016): 50–53.

Lauer, W. C., L. W. Condie, G. W. Wolfe, E. T. Czeh, and J. M. Burns. "Denver Potable Water Reuse Demonstration Project: Comprehensive Chronic Rat Study." *Food and Chemical Toxicology* 32, no. 11 (1994): 1021–30.

Law, I. B. "Advanced Reuse—from Windhoek to Singapore and Beyond." *Water* 30, no. 5 (2003): 44–50.

Lesser, L. E., A. Mora, C. Moreau, J. Mahlknecht, A. Hernández-Antonio, A. I. Ramírez, and H. Barrios-Piña. "Survey of 218 Organic Contaminants in Groundwater Derived from the World's Largest Untreated Wastewater Irrigation System: Mezquital Valley, Mexico." *Chemosphere* 198 (May 2018): 510–21.

Leverenz, H. L., G. Tchobanoglous, and T. Asano. "Direct Potable Reuse: A Future Imperative." *Journal of Water Reuse and Desalination* 1, no. 1 (2011): 2–10.

Limerick, P. N., and J. L. Hanson. *A Ditch in Time*. Golden, CO: Fulcrum, 2012.

Livingston, P. P., and R. R. Bennett. *Geology and Ground-Water Resources of the Big Spring Area, Texas*. Washington, DC: U.S. Geological Survey, 1944.

Mac Kenzie, W. R., N. J. Hoxie, M. E. Proctor, M. S. Gradus, K. A. Blair, D. E. Peterson, J. J. Kazmierczak, et al. "A Massive Outbreak in Milwaukee of *Cryptosporidium* Infection Transmitted through the Public Water Supply." *New England Journal of Medicine* 331, no. 3 (1994): 161–67.

Marella, R. L. *Water Withdrawals, Use, and Trends in Florida, 2015*. Reston, VA: U.S. Geological Survey, 2020.

Medema, G., L. Heijnen, G. Elsinga, R. Italiaander, and A. Brouwer. "Presence of SARS-Coronavirus-2 RNA in Sewage and Correlation with Reported

COVID-19 Prevalence in the Early Stage of the Epidemic in the Netherlands." *Environmental Science & Technology Letters* 7, no. 7 (2020): 511–16.

Meehan, K., K. J. Ormerod, and S. A. Moore. "Remaking Waste as Water: The Governance of Recycled Effluent for Potable Water Supply." *Water Alternatives* 6, no. 1 (2013): 67–85.

Megdal, S. B., P. Dillon, and K. Seasholes. "Water Banks: Using Managed Aquifer Recharge to Meet Water Policy Objectives." *Water* 6 (2014): 1500–14.

Mehan, G. T. "Found Water: Reuse and the Deconstruction of 'Wastewater.'" *Water Finance & Management*, December 10, 2019.

Metzler, D. F., R. L. Culp, H. A. Stoltenberg, R. L. Woodward, G. Walton, S. L. Chang, N. A. Clarke, C. M. Palmer, and F. M. Middleton. "Emergency Use of Reclaimed Water for Potable Supply at Chanute, Kan." *Journal AWWA* 50, no. 8 (1958): 1021–60.

Muller, D. A., and R. D. Price, *Ground-Water Availability in Texas—Estimates and Projections through 2030*. Austin: Texas Department of Water Resources, 1979.

Nading, T., L. Schimmoller, D. Holloway, T. Henifin, J. Dano, G. Salazar-Benites, C. Wilson, C. Bott, and J. Mitchell. "A 'SWIFT' Approach to Managed Aquifer Recharge." *Water Online*, January 24, 2018. www.wateronline.com/doc/a-swift-approach-to-managed-aquifer-recharge-0001.

National Academies of Sciences, Engineering, and Medicine. *Using Graywater and Stormwater to Enhance Local Water Supplies: An Assessment of Risks, Costs, and Benefits*. Washington, DC: National Academies Press, 2016.

National Research Council. *Ground Water Recharge Using Waters of Impaired Quality*. Washington, DC: National Academy Press, 1994.

———. *Issues in Potable Reuse: The Viability of Augmenting Drinking Water Supplies with Reclaimed Water*. Washington, DC: National Academy Press, 1998.

———. *Quality Criteria for Water Reuse*. Washington, DC: National Academy Press, 1982.

———. *Water Reuse: Potential for Expanding the Nation's Water Supply through Reuse of Municipal Wastewater*. Washington, DC: National Academies Press, 2012.

National Water Research Institute. *Developing Proposed Direct Potable Reuse Operational Procedures and Guidelines for Cloudcroft, New Mexico*. Fountain Valley, CA: National Water Research Institute, 2015.

———. *Enhanced Source Control Recommendations for Direct Potable Reuse in California*. Fountain Valley: Report prepared for California State Water Resources Control Board, 2020. Available at www.waterboards.ca.gov/drinking_water/certlic/drinkingwater/docs/dpr-esc-2020.pdf.

Nix, D., M. Southard, H. Burris, H. Adams, and R. Schreiber. "Tracking Wichita Falls' Path from DPR to IPR." *Opflow* 47, no. 2 (2021): 10–15.

Okun, D. A. "Water Reclamation and Unrestricted Nonpotable Reuse: A New Tool in Urban Water Management." *Annual Review of Public Health* 21 (2000): 223–45.

Olivieri, A. W., J. Crook, M. A. Anderson, R. J. Bull, J. E. Drewes, C. N. Haas, W. Jakubowski, et al. *Expert Panel Final Report: Evaluation of the Feasibility of Developing Uniform Water Recycling Criteria for Direct Potable Reuse.* Sacramento: Prepared by the National Water Research Institute for the State Water Resources Control Board, 2016.

Ongerth, H. J., and J. E. Ongerth. "Health Consequences of Wastewater Reuse." *Annual Reviews in Public Health* 3 (1982): 419–44.

Orange County Sanitation District. *2019–2020 Pretreatment Program Annual Report.* Available www.ocsd.com/home/showpublisheddocument?id=30137.

Orange County Water District and Orange County Sanitation District. *GWRS: Groundwater Replenishment System.* Available at www.ocwd.com/media/8861/ocwd-technicalbrochure_web-2020.pdf.

Ormerod, K. J., and L. Silvia. "Newspaper Coverage of Potable Water Recycling at Orange County Water District's Groundwater Replenishment System, 2000–2016." *Water* 9, no. 12 (2017): 984. doi.org/10.3390/w9120984.

Parker, A. R., and C. R. Lawrence. "Water Capture by a Desert Beetle." *Nature* 414, no. 6859 (2001): 33–34.

Paschke, S. S., ed. *Groundwater Availability of the Denver Basin Aquifer System, Colorado.* Reston, VA: U.S. Geological Survey, 2011.

Patterson, L. A., and M. W. Doyle. *2020 Aspen-Nicholas Water Forum Water Affordability and Equity Briefing Document.* Durham, NC: Nicholas Institute, 2020.

Pecson, B., D. Gerrity, K. Bibby, J. E. Drewes, C. Gerba, R. Gersberg, R. Gonzalez, et al. "Editorial Perspectives: Will SARS-CoV-2 Reset Public Health Requirements in the Water Industry? Integrating Lessons of the Past and Emerging Research." *Environmental Science: Water Research & Technology* 6, no. 7 (2020): 1761–64.

Pecson, B. M., R. S. Trussell, A. N. Pisarenko, and R. R. Trussell. "Achieving Reliability in Potable Reuse: The Four Rs." *Journal AWWA* 107, no. 3 (2015): 48–58.

Pruden, A. "Balancing Water Sustainability and Public Health Goals in the Face of Growing Concerns about Antibiotic Resistance." *Environmental Science & Technology* 48, no. 1 (2014): 5–14.

Pyne, R. D. G. *Aquifer Storage Recovery*, 2nd ed. Gainesville, FL: ASR Press, 2005.

Qadir, M., G. C. Jiménez, R. L. Farnum, and P. Trautwein. "Research History and Functional Systems of Fog Water Harvesting." *Frontiers in Water* 3 (2021). https://doi.org/10.3389/frwa.2021.675269.

Quartiano, M. *Toilet to Tap: San Diego's "Pipe Dream."* Self-published manuscript, 2006. Available at San Diego Public Library.

Rand, H. *Water Wars: A Story of People, Politics and Power.* Philadelphia: Xlibris, 2003.

Reuse Coordinating Committee and the Water Conservation Initiative Water Reuse Work Group. *Water Reuse for Florida: Strategies for Effective Use of Reclaimed Water.* Tallahassee: Florida Department of Environmental Protection, 2003. Available at https://floridadep.gov/sites/default/files/valued_resource_FinalReport_508C.pdf.

Richter, B. D., K. Benoit, J. Dugan, G. Getacho, N. LaRoe, B. Moro, T. Rynne, M. Tahamtani, and A. Townsend. "Decoupling Urban Water Use and Growth in Response to Water Scarcity." *Water* 12, no. 10 (2020): 1–14. www.mdpi.com/2073-4441/12/10/2868.

Rose, J. B., D. E. Huffman, K. Riley, S. R. Farrah, J. O. Lukasik, and C. L. Hamann. "Reduction of Enteric Microorganisms at the Upper Occoquan Sewage Authority Water Reclamation Plant." *Water Environment Research* 73, no. 6 (2001): 711–20.

Rozin, P., B. Haddad, C. Nemeroff, and P. Slovic. "Psychological Aspects of the Rejection of Recycled Water: Contamination, Purification and Disgust." *Judgment and Decision Making* 10, no. 1 (2015): 50–63.

San Francisco Public Utilities Commission. *Onsite Water Reuse Program Guidebook.* San Francisco: San Francisco Public Utilities Commission, 2020. Available at https://sfpuc.org/sites/default/files/documents/Onsite WaterReuseGuidebook2020.pdf.

Sauvé, S., and M. Desrosiers. "A Review of What Is an Emerging Contaminant." *Chemistry Central Journal* 8 (2014): 15. doi.org/10.1186/1752-153X-8-15.

Schwabe, K., M. Nemati, R. Amin, Q. Tran, and D. Jassby. "Unintended Consequences of Water Conservation on the Use of Treated Municipal Wastewater." *Nature Sustainability* 3 (2020): 628–35.

Scott, T. M., G. H. Means, R. P. Meegan, R. C. Means, S. B. Upchurch, R. E. Copeland, J. Jones, T. Roberts, and A. Willet. *Springs of Florida.* Tallahassee: Florida Geological Survey, 2004.

Scruggs, C. E., and B. M. Thomson. "Opportunities and Challenges for Direct Potable Water Reuse in Arid Inland Communities." *Journal of Water Resources Planning and Management* 143, no. 10 (2017): 04017064.

Scruggs, C. E., C. B. Pratesi, and J. R. Fleck. "Direct Potable Water Reuse in Five Arid Inland Communities: An Analysis of Factors Influencing Public Acceptance." *Journal of Environmental Planning and Management* 63, no. 8 (2019): 1470–1500.

Scruggs, C. E., D. F. Lawler, G. Tchobanoglous, B. M. Thomson, M. R. Schwarzman, K. J. Howe, and A. J. Schuler. "Potable Water Reuse in Small Inland Communities: Oasis or Mirage?" *Journal AWWA* 112, no. 4 (2020): 10–17.

Sedlak, D. *Water 4.0: The Past, Present, and Future of the World's Most Vital Resource*. New Haven, CT: Yale University Press, 2014.

Sharvelle, S., L. A. Roesner, Y. Qian, M. Stromberger, and M. N. Azar. *Long-Term Study on Landscape Irrigation Using Household Graywater—Experimental Study*. Fort Collins, CO: Report to Water Environment & Reuse Foundation, 2012.

Sharvelle, S., N. Ashbolt, E. Clerico, R. Hultquist, H. Leverenz, and A. Olivieri. *Risk-Based Framework for the Development of Public Health Guidance for Decentralized Non-Potable Water Systems*. Alexandria, VA: Prepared by the National Water Research Institute for the Water Environment & Reuse Foundation, 2017.

Shin, C., P. L. McCarty, J. Kim, and J. Bae. "Pilot-Scale Temperate-Climate Treatment of Domestic Wastewater with a Staged Anaerobic Fluidized Membrane Bioreactor (SAF-MBR)." *Bioresource Technology* 159 (2014): 95–103.

Siegel, S. M. *Troubled Water: What's Wrong with What We Drink*. New York: Thomas Dunne, 2019.

Slayton, R. "Sludge Busters." *Popular Science* (February 1987): 43–44.

Southwest Florida Water Management District. "50 Years." *WaterMatters Magazine* (October 2011). Available at www.swfwmd.state.fl.us/sites/default/files/medias/documents/50thAnniversary-WaterMatters.pdf.

Steinle-Darling, E. "Total Water Solutions—the Many Faces of DPR in Texas." *Journal AWWA* (March 2015): 16–20.

Stevens, L. A. *The Town That Launders Its Water*. New York: Coward, McCann & Geoghegan, 1971.

Taleb, N. N. *The Black Swan: The Impact of the Highly Improbable*. New York: Random House, 2007.

Tchobanoglous, G., J. Cotruvo, J. Crook, E. McDonald, A. Olivieri, A. Salveson, and R. S. Trussell. *Framework for Direct Potable Reuse*. Alexandria, VA: WateReuse Research Foundation, 2015.

Teodoro, M. P., and R. R. Saywitz. "Water and Sewer Affordability in the United States: A 2019 Update." *AWWA Water Science* 2, no. 2 (2020): e1176.

Ternes, T. A. "Occurrence of Drugs in German Sewage Treatment Plants and Rivers." *Water Research* 32, no. 11 (1998): 3245–60.

Udall, B., D. Kenney, and J. Fleck. "Western States Buy Time with a 7-Year Colorado River Drought Plan, But Face a Hotter, Drier Future." *The Conversation*, July 10, 2019, https://theconversation.com/

western-states-buy-time-with-a-7-year-colorado-river-drought-plan-but-face
-a-hotter-drier-future-119448.

USDA Forest Service. *Water and the Forest Service*. Washington, DC: USDA
Forest Service, 2000.

U.S. Environmental Protection Agency. *2017 Potable Reuse Compendium*.
Washington, DC: U.S. Environmental Protection Agency, 2017.

———. *Clean Watersheds Needs Survey 2012 Report to Congress*. Washington, DC: U.S. Environmental Protection Agency, 2016.

———. *Guidelines for Water Reuse*. Washington, DC: U.S. Environmental
Protection Agency, 2012.

———. *Mainstreaming Potable Water Reuse in the United States: Strategies
for Leveling the Playing Field*. ReNUWIt and the Johnson Foundation, 2018.

———. *National Water Reuse Action Plan: Collaborative Implementation
(Version 1)*. Washington, DC: U.S. Environmental Protection Agency, 2020.
Available at www.epa.gov/waterreuse/national-water-reuse-action-plan
-collaborative-implementation-version-1.

———. *National Water Reuse Action Plan Draft*. Washington, DC: U.S.
Environmental Protection Agency, 2019. Available at www.epa.gov/sites/
production/files/2019-09/documents/water-reuse-action-plan-draft-2019
.pdf.

Walker, D. *Thirst for Independence: The San Diego Water Story*. San Diego:
Sunbelt Publications, 2004.

Webber, M. E. *Thirst for Power*. New Haven, CT: Yale University Press, 2016.

Western Consortium for Public Health. *The City of San Diego Total Resource
Recovery Project: Health Effects Study—Final Summary Report*. Oakland,
CA: Western Consortium for Public Health, 1992.

William J. Worthen Foundation. *A Design Professional's Guide to Onsite
Water Use and Reuse*. San Francisco: William J. Worthen Foundation, 2018.

Yu, Z. L. T., A Rahardianto, J. R. DeShazo, M. K. Stenstrom, and Y. Cohen.
"Critical Review: Regulatory Incentives and Impediments for Onsite Graywater Reuse in the United States." *Water Environment Research* 85, no. 7
(2013): 650–62.

Index

advanced oxidation, xiv, 40, 53, 67,
107, 150;
1,4-dioxane and, 43
agricultural irrigation, 14, 36, 60, 107;
Florida, 89–90;
Monterey, 18–19, 168-69;
sewer farms, 17
air conditioning condensate,
recycling of, 156–57, 164
Alabama, 76–77, 92
American Water Works Association,
15, 165, 170
amount of water reuse, xi, 107, 169;
in Florida, 90, 93
anaerobic processes, 165–66
antibiotic-resistant bacteria, 118
Arizona:
Arizona Pure Water Brew, 127;
Arizona Water Resources
Research Center, 112;
Colorado River and, 31, 37, 112;
direct potable reuse and, 114, 152;
Scottsdale, 111–14
Aurora, Colorado:
early water reuse, 65;
Prairie Waters Project, 66-68;

water-supply issues, 65–66;
WISE partnership and, 69, 71–72
Austin, Texas, 156, 164-65
Australia:
sewer farms, 17;
Toowoomba, 135–37;
Western Corridor Recycled Water
Scheme, 136

bacteria, 2, 84, 104–5, 108, 124, 131,
135, 158;
personal care products and, 117;
water treatment and, xiii, 4, 39,
166.
See also antibiotic-resistant
bacteria
Barnett, Cynthia, 91
beer:
beer making as public relations
tool, 126–27;
reuse of brewery process water,
163
Belanger, Laura, 66
Biesemeyer, Brian, 114
Big Spring, Texas, 143–44, 152
bioassays, 5, 119

219

About the Authors

Dr. William (Bill) M. Alley is an internationally recognized authority on groundwater and water quality. He was chief of the Office of Groundwater for the U.S. Geological Survey for almost two decades and serves as director of science and technology for the National Ground Water Association.

Rosemarie Alley is a freelance writer with extensive writing and public speaking experience. As a writing team, Bill provides the scientific expertise, and Rosemarie makes it interesting and understandable for the general reader. Bill and Rosemarie previously collaborated on *Too Hot to Touch: The Problem of High-Level Nuclear Waste* (2013), *High and Dry: Meeting the Challenges of the World's Growing Dependence on Groundwater* (2017), and *The War on the EPA: America's Endangered Environmental Protections* (2020). The Alleys live in San Diego, California.